Engineering the Human

Bert-Jaap Koops · Christoph H. Lüthy
Annemiek Nelis · Carla Sieburgh
J. P. M. Jansen · Monika S. Schmid
Editors

Engineering the Human

Human Enhancement Between Fiction and Fascination

Editors

Bert-Jaap Koops
Tilburg Institute for Law,
 Technology and Science
Tilburg University
Tilburg
The Netherlands

Christoph H. Lüthy
Faculty FTR
Center for the History of Philosophy
 and Science
Radboud University Nijmegen
Nijmegen
The Netherlands

Annemiek Nelis
Dutch Safety Board
The Hague
The Netherlands

Carla Sieburgh
Faculty of Law
Radboud University Nijmegen
Nijmegen
The Netherlands

J. P. M. Jansen
Department of English Linguistics
Faculty of Arts
University of Groningen
Groningen
The Netherlands

Monika S. Schmid
Department of English Linguistics
Faculty of Arts
University of Groningen
Groningen
The Netherlands

ISBN 978-3-642-35095-5   ISBN 978-3-642-35096-2   (eBook)
DOI 10.1007/978-3-642-35096-2
Springer Heidelberg New York Dordrecht London

Library of Congress Control Number: 2012955747

© Springer-Verlag Berlin Heidelberg 2013
This work is subject to copyright. All rights are reserved by the Publisher, whether the whole or part of the material is concerned, specifically the rights of translation, reprinting, reuse of illustrations, recitation, broadcasting, reproduction on microfilms or in any other physical way, and transmission or information storage and retrieval, electronic adaptation, computer software, or by similar or dissimilar methodology now known or hereafter developed. Exempted from this legal reservation are brief excerpts in connection with reviews or scholarly analysis or material supplied specifically for the purpose of being entered and executed on a computer system, for exclusive use by the purchaser of the work. Duplication of this publication or parts thereof is permitted only under the provisions of the Copyright Law of the Publisher's location, in its current version, and permission for use must always be obtained from Springer. Permissions for use may be obtained through RightsLink at the Copyright Clearance Center. Violations are liable to prosecution under the respective Copyright Law.
The use of general descriptive names, registered names, trademarks, service marks, etc. in this publication does not imply, even in the absence of a specific statement, that such names are exempt from the relevant protective laws and regulations and therefore free for general use.
While the advice and information in this book are believed to be true and accurate at the date of publication, neither the authors nor the editors nor the publisher can accept any legal responsibility for any errors or omissions that may be made. The publisher makes no warranty, express or implied, with respect to the material contained herein.

Printed on acid-free paper

Springer is part of Springer Science+Business Media (www.springer.com)

Contents

1 Towards Homo Manufactus? An Introduction to this Volume . . .   1
Christoph H. Lüthy and Bert-Jaap Koops

2 Historical and Philosophical Reflections on Natural, Enhanced and Artificial Men and Women   11
Christoph H. Lüthy

3 Changing the Body Through the Centuries   29
Theo Mulder

4 Human Enhancement in Futures Explorations   45
Lucas Cornips and Marjolein van Asselt

5 Genetic Enhancement of Human Beings: Reality or Fiction?   63
Annemiek Nelis and Danielle Posthuma

6 Gulliver's Next Travels: A Journey into the Land of Biomaterials and Synthetic Life   71
Annemiek Nelis and Hub Zwart

7 Human Robots and Robotic Humans   83
Catholijn M. Jonker and Annemiek Nelis

8 Human Enhancement, Evolution and Lifespan: Evolving Towards Immortality?   101
Simon Verhulst

9 Opting for Prevention: Human Enhancement and Genetic Testing   113
Annemiek Nelis, Symone Detmar and Elske van den Akker

10 A Unique Copy: The Life and Identity of Clones in Literary Fiction 129
Bert-Jaap Koops

11 Parents' Responsibility for Their Choices Regarding the Enhancement of Their Child 151
Carla Sieburgh

12 Concerning 'Humans' and 'Human' Rights. Human Enhancement from the Perspective of Fundamental Rights 165
Bert-Jaap Koops

13 Conclusion: The Debate About Human Enhancement 183
Bert-Jaap Koops

Index 193

Chapter 1
Towards Homo Manufactus?
An Introduction to this Volume

Christoph H. Lüthy and Bert-Jaap Koops

Abstract This contribution explores how the concept of human engineering emerged and what place it assumes in contemporary debate. The term has recently been used in discussions on a range of subjects, among which are technology, science and sports. As the number of different ways of adjusting the human body keeps growing, the idea of 'transhumans' is taking hold in today's society. Although scientists generally consider it unlikely that 'transhumans' will become a reality in the foreseeable future, the concept still causes fear, raises hopes and leads to numerous questions. The main issue is whether or not it is ethical to interfere with the human body to such an extent. While it is certain that these kinds of changes can transform the human condition, the extent to which this is possible remains unclear.

Transhumanist Scenarios

The Transhumanist Declaration of 1998 begins with the following statement:

> §1. Humanity stands to be profoundly affected by science and technology in the future. We envision the possibility of broadening human potential by overcoming 'aging', cognitive

---

C.H. Lüthy (✉)
Faculty FTR, Center for the History of Philosophy and Science,
Radboud University Nijmegen, Erasmusplein 1, 6525 HT, Nijmegen, The Netherlands
e-mail: c.luethy@ftr.ru.nl
URL: http://www.ru.nl/philosophy/chps/current_members/luthy/

B.-J. Koops
Tilburg Institute for Law, Technology and Science, Tilburg University,
P.O. Box 90153, 5000 LE, Tilburg, The Netherlands
e-mail: E.J.Koops@uvt.nl
URL: http://www.tilburguniversity.edu/webwijs/show/?uid=e.j.koops

B.-J. Koops et al. (eds.), *Engineering the Human*, DOI: 10.1007/978-3-642-35096-2_1,
© Springer-Verlag Berlin Heidelberg 2013

shortcomings, involuntary suffering, and our confinement to planet Earth (The Transhumanist Declaration 1998).

One is tempted to reformulate these sentences in the present tense: 'today, humanity is profoundly affected by science and technology'. Does our rising life expectancy not testify to impressive successes in combating the process of ageing? Are our cognitive shortcomings not already made up for by electronic gadgets and psychopharmaceuticals? Is much involuntary suffering not being alleviated or entirely done away by today's medical treatment?

The Transhumanist Declaration (1998) is, however, not about recent medical, technological and scientific advances, but emphasises a vision of the near future—a prediction, moreover, which it welcomes and embraces. It is based on the assumption that the various recent technological accomplishments will soon converge, and that this convergence should bring about a new type of human being, the 'transhuman' mentioned in the manifesto's title.

A number of scenarios have been developed, some by real or would-be scientists, others by science fiction authors or filmmakers, in which the world resembles that of *Star Wars*, where human beings live together with intelligent robots and modified man-machines. 'These will soon become symbiotic, leading to a synergy between men and machines that few anticipated', according to Benford and Malartre (2007) (196). Warwick (2003), for one, of the Department of Cybernetics at the University of Reading, is convinced that 'the era of the Cyborg is now upon us', the Cyborg being 'part human part machine' (131). The inventor and science author Kurzweil (2005, 2006), in turn, predicts that

> the most important and radical application particularly of circa-2030 nanobots will be to expand our minds through the merger of biological and nonbiological or machine intelligence. In the next 25 years, we will learn how to augment our 1000 trillion very slow interneuronal connections with highspeed virtual connections via nanorobotics. This will allow us to greatly boost our pattern-recognition abilities, memories, and overall thinking capacity, as well as to directly interface with powerful forms of computer intelligence. The technology will also provide wireless communication from one brain to another. In other words, the age of telepathic communication is almost upon us. (43)

Or take the philosopher Bostrom (2003) at Oxford University, who in 2003 announced that he was preparing himself ethically for our future as 'transhumans', that is, genetically and bionically modified creatures that Bostrom (2003) hopes will be 'healthier, wittier, happier people', who moreover 'may be able to reach new levels culturally' (498).

The majority of contemporary scientists find most of these predictions highly unrealistic. They either consider it unlikely that the envisaged merger of nanotechnology, engineering and biotechnology can be carried out as predicted; or they reject the proposed time frame between 2020 and 2050 as implausibly soon; or, when they do give some credit to these scenarios, they suggest that legislation or ethical standards will prevent them from being implemented.

Man-made Man?

Whether plausible or not, such scenarios inevitably provoke discussions, cause anxieties, engender fantasies and nurture expectations. Discussion may take on a variety of forms, ranging from science fiction novels and movies to proceedings of ethics conferences, from advisory policy reports to public debates. Moreover, each country or, rather, each linguistic community conducts these discussions differently. This has to do with the terminology that is used to refer to the bundle of medical, technological and scientific procedures that are allegedly transforming humankind. In English, the term 'human enhancement' dominates the debate, implying the improvement of the already existing functions and capacities, while the alternative terms 'artificial man' or 'transhuman' imply a disruptive discontinuity between current, naturally engendered human forms and future, artificial ones. The German expression 'die Perfektionierung des Menschen' ('perfectioning of man'), by contrast, possesses, like 'enhancement', a positive connotation of improvement, but not of discontinuity. The alliterative Dutch expression 'de maakbare mens' ('makeable man'), in turn, provides a more value-neutral term that can include any of the current techniques applied to changing human nature— not all of which need to aim at enhancement.

The present collection of essays was first written for a Dutch-speaking audience, and it carried in its original title the local catch-all term—'makeable man'— which indeed stands for all kinds of procedures enhancing, improving or indeed engineering humans. The 12 sections of the 2003 Technology Festival held at Amsterdam, which dealt with the issue of the 'makeable man', convey an idea of the diverse connotations of this term:

1. Cloning
2. Prenatal selection of babies
3. Gene therapy
4. Techniques of conditioning behaviour
5. Neurosurgery
6. Replacement medicine
7. Cosmetic surgery
8. Anti-ageing
9. Top-class sport (enhanced performance)
10. Cybernetics (applying artificial intelligence to human beings)
11. Nanotechnology and its use inside the human body
12. Nutrition

It turns out that this untranslatable catch-all term, 'makeable man', offers a range of advantages over expressions such as 'human enhancement'. Precisely because of the all-inclusiveness of the term, Dutch and Flemish society has benefited from a comprehensive discussion. The debate has taken future scenarios of converging technological, medical and scientific advances seriously, has attempted to gauge their likelihood and to fathom possible advantages and disadvantages, and has contemplated the ethical and political limits that ought possibly to be formulated. Here are some examples. The just-named 2003 Technology Festival in Amsterdam was entitled 'Homo Sapiens 2.0: Festival about the "Makeable Man"'. In 2004, the Flemish Institute for Science and Technology Assessment organised an essay contest with 'Makeable Man' as its theme. In translation, the description of the essay question read as follows: 'Artificial muscles for the disabled. A chip implanted in your head. Technology makes man. Dream or nightmare?' Three years later, in 2007, the Rathenau Institute, a technology assessment body advising Dutch parliament, asked scientists and philosophers whether there should be limits to the engineering of 'makeable humans'. Yet another year later, an organisation called 'Makeable Man' (*De Maakbare Mens*), which describes itself as a 'critical movement for bio-ethics', invited entries for a photo contest about 'Sports and the makeability of humans' (www.demaakbaremens.org). Finally, Maastricht University has over the past few years offered its students a course entitled 'Makeable Man' in its Bachelor degree programme 'Arts and Culture'. This list could be continued ad nauseam; for example, by adding numerous magazine and newspaper articles that have addressed the issue.

The question is warranted whether a debate that covers such a broad range of heterogeneous practices can possibly be meaningful. Will it not necessarily mix up separate issues in a general scenario that, however unrealistic, is likely to engender only fear? The illustration on the programme flyer of the 'Homo Sapiens 2.0' festival displayed plastic mannequins, in a gesture towards a future in which human beings will be artificially produced that bear only a superficial resemblance to the humans they replace. The cover of the syllabus of Maastricht's bachelor course (Fig. 1.1) shows a picture of a drawer divided into many small compartments, which are filled with human heads, conjuring up the idea of a repository in which the engineers of humanity can store spare parts and from which, whenever needed, a replacement head can be taken out. In short, then, the suggestion is invoked that it will soon be possible to reform, perfect, standardise or indeed replace 'naturally evolved' human beings by engineered specimens. Since such a wholesale replacement presently belongs to the realm of fiction, not of fact, one may in fact wonder about the usefulness of such scenarios. Is it helpful to lump cloning, conditioned behaviour, anti-ageing techniques, cosmetic surgery and performance-enhancing drugs together and view them as so many stepping stones on our way towards the creation of artificial life? It could perhaps be more meaningful to highlight the generic differences, rather than stretching some similarities, between the following types of interventions: (1) enhancement of the existing functions; (2) methods of selection in the reproduction of human individuals and possible improvements of the genetic makeup of the embryo; (3) replacement or expansion of natural elements by artificial elements (from replacing organs to the creation of cyborgs); (4) methods designed to steer human behaviour; (5) the development of robots that increasingly resemble humans.

1 Towards Homo Manufactus? An Introduction to this Volume

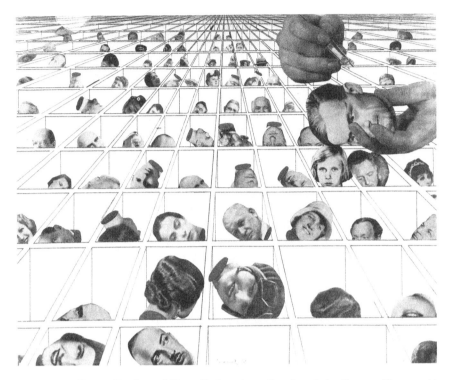

Fig. 1.1 Kurt Kranz, 'Kopfvorrat'. From: Barbara Auer, *Künstler mit der Kamera. Photographie als Experiment*, Mannheim, Vits and Kehrer 1994.

What Lies at the Core of the Debate?

The main reason for asking academics from a variety of disciplines, ranging from reproductive biology over artificial intelligence and law to the history of science, to reflect on 'makeable man' was exactly so as to gauge the coherence of the debate. When viewed from a number of scientific perspectives, do short- and long-term projections of technological, medical and scientific developments justify such a general and as yet hypothetical debate? Or is it driven by merely fictional scenarios that do not accomplish more than to give rise to unfounded hopes and fears and provoke distressingly unanswerable questions? Is 'the future of our selves' really at risk, as was suggested in the title of a 2002 advisory report from the Health Council of the Netherlands?

There are the optimists, cited at the beginning of this Introduction, who welcome the convergence of various human-related technologies in the coming decades and the advent of superman. Among the optimists are not only pioneering scientists such as Warwick (2003) or visionaries like Kurzweil (2005, 2006) but also ethicists such as Harris (2007) who claim a moral duty to enhance ourselves. Still on the optimistic side of the spectrum are those who, like lawyer Gavaghan (2007) in

Defending the Genetic Supermarket, use rational argumentation to challenge many fallacies used in the debate. Gavaghan (2007) argues that—barring really harmful consequences which are seldom proven to be realistic—individuals have the right to decide for themselves whether or not, and how, to engineer human life.

In the middle of the spectrum, we find those who think that it is our moral task rather to be conceptually ready with nuanced answers for all eventualities, irrespective of whether any of the prophecies will come true. This is the position, for example, of the German ethicist Gesang (2007) whose recent survey book, *Die Perfektionierung des Menschen*, attempts to find utilitarian answers to questions regarding the demarcation between desirable, and hence permissible, and undesirable and illegitimate alterations of human nature. The Oxford ethicist Glover (1984), in turn, tries to answer the question: *What Sort of People Should There Be?* He argues that there are certain aspects to human nature which might become stronger with the help of biotechnology rather than being threatened by it. The breadth of the middle ground is illustrated by the many contributions to the volume *Human Enhancement*. Its editors, Savulescu and Bostrom (2009), pp. 18–19, stress that the issue has moved from the realm of fiction to that of practical ethics. This implies that part of the debate should now focus on the specifics of disaggregated forms of enhancement, while another part needs to develop a long-term and big-picture perspective on the future of humanity.

At the pessimistic end of the spectrum, we find those who warn against the de-humanisation of humans. Kass (2002), chairman of the President's Council on Bioethics under the former US president George W. Bush, emphasises the fact that all important aspects of human life—including work, sexuality, food, rituals—are meaningless when they are placed outside of our traditional life cycle. In order to preserve meaning, we must, therefore, preserve this cycle from birth to death. For similar reasons, Fukuyama (2002) argues that human life will lose its meaning if we design out human suffering and bad luck altogether. After all, happiness is only possible if people know the meaning of unhappiness as well. Therefore, he considers the proposal to raise human beings to a new level with the help of bio-technology to be 'the world's most dangerous idea'. Dekker (2007), professor and molecular biophysicist at Delft University of Technology, agrees with Fukuyama (2002, 2004): 'This might sound like a drastic statement, but I agree with it'. After all, he continues, 'I support technology's commitment to heal human beings, but I object against the endeavour to improve humans out of a sense of hubris, which will lead to dangerous side effects'. Of these side effects, the most dangerous is, according to Dekker (2007), the loss of 'human dignity'. In a similar vein, Sandel (2007), pp. 96–97 in *The Case Against Perfection*, warns against the consequences of losing our 'openness to the unbidden' in engineering human life: we will no longer value natural gifts or show humility in the face of privilege, and we may lose the knack of improving the world around us if all we do is try to improve ourselves.

The pessimists do not tire of warning us of the impending loss of 'naturalness'. Even though it might be argued that human beings were driven from the paradise of 'naturalness' long ago, the current impression of a potential loss of naturalness seems caused by the speed and intensity of progress in, particularly, the bio-technological domain. History shows that public debates are usually not caused by changes themselves, but rather by the speed at which these changes take place. Changes which happen slowly and almost imperceptibly tend to cause little resistance, while changes happening so fast that they become noticeable often incite public debate. As the historian Bess (2008) mentions in his study about the history of biological 'enhancement':

> This time around, however, the radical innovations are coming upon us suddenly, in a matter of decades. Contemporary society is unprepared for the dramatic and destabilizing changes it is about to experience, down this road on which it is already advancing at an accelerating pace.

Indeed, we live in a time of rapid technological innovations, not least in the biomedical field, which are often publicly financed and affect all areas of our lives. These changes are reflected in life statistics: we see a decrease in the number of infant deaths through the prevention of infections, and also an equally strong increase of average life expectancy. Economically, these changes are paired with a noticeable and still increasing use of the medical sciences. Culturally, they are expressed through the flourishing of a health culture and the glorification of 'body consciousness', an awareness of the body in general and our own, individual bodies in particular. Scientifically, these changes are both represented and pushed forward by an ever growing group of scientists and professionals in the life sciences.

It is precisely this conjunction of far-reaching changes in healthcare with achievements in such sciences as robotics and artificial intelligence that lies behind the diffuse but widespread fantasies of man-made man, of the artificially improved, eternally healthy and possibly immortal human being. It is the just-named com-bination of developments that endows the prophecies of the post-human *Über-mensch* with a certain degree of credibility. After all, artificial intelligence, pre-implantation genetic diagnostics, genetic engineering, drugs for the improvement of memory, concentration, alertness and mood, together with performance enhancers, plastic surgery, sex-change operations, prosthetic medicine, anti-ageing medicine and direct interaction between human beings and machines—these are all types of technological interventions that are already existent today, and they are starting to be combined in remarkable, often unimagined manners.

Between Fiction and Fascination

But again, how realistic are the different ideas of the future with which futuro-logically inclined scientists or visionaries confront us? What are the actual sci-entific and technological possibilities, and how will they further develop? What are the chances that current fictional ideas, based on the predictions of both futurists and pessimists, will become reality? Also, if we assume that some of these ideas will become real, what will be the implications for society and individuals? These are questions to which the authors of this book have been asked to respond.

They have done so in a number of ways. One obvious way in which to address such a cluster of questions is by placing it in a historical perspective. Such a method manages to show that a good portion of our expectations and fears has a long history and that our concerns lose some of their urgency and seeming novelty when placed in a broader historical perspective. We need only mention Rostand's (1959) *Can Man Be Modified?* and Ramsey's (1970) *Fabricated Man* here to show that the participants in today's human enhancement debate are hardly discussing radically new issues. This method is used in some of the initial essays of this book. Some of the other contributions document, by contrast, that professionals who are actually working in fields that shape our human future do not consider the scenarios sketched by the prophets of human engineering to be realistic. A further method for dealing with this cluster of issues is the traditional approach of dividing the general discussion into thematic issues, such as prenatal diagnostics, artificial intelligence or human rights. Such an approach helps us understand that the supposed whole is bigger than the sum of the separate parts, but also that the individual parts are easier to understand on their own. This method is used either implicitly or explicitly by most authors in this book.

This book originated from the decision of a handful of members of The Young Academy (*De Jonge Akademie*), the younger sister of the Royal Netherlands Academy of Arts and Sciences, in collaboration with the Centre for Society and Genomics (CSG) of Radboud University Nijmegen, to get a better understanding of the ongoing debate surrounding the enhancement of humans and their possible transformation into post-humans. By combining the perspectives of many different disciplines, the authors hope to enhance (but not engineer) the international debate on 'makeable man'.[1]

Acknowledgments The papers in this collection were translated from the original Dutch by students of the MA Writing, Editing and Mediating at the University of Groningen. Publication of this volume was made possible through the generous financial support of The Young Academy (De Jonge Akademie). We are very grateful to Lydia ten Brummelhuis for her careful and dedicated work in editing the final manuscript.

---

[1] The Young Academy, established in 2005, counts 50 members, which have been selected for their academic excellence and international reputation, as well as their interdisciplinary background and methodology. Since The Young Academy includes representatives from all academic disciplines, the topic of this book appeared to be an ideal topic for a collective research project. The book project was made possible through a fruitful cooperation with the Centre for Society and Genomics (CSG) at Radboud University Nijmegen, which investigates the relationship between society and genomics.

Bibliography

Benford G, Malartre E (2007) Beyond human: living with robots and cyborgs. Tom Doherty Associates, New York

Bess M (2008) Icarus 2.0: A historian's perspective on human biological enhancement. Techno Cult 49(1):114–126

Bostrom N (2003) Human genetic enhancements: a transhumanist perspective. J Value Inq 37:493–506

Bostrom N (2005) A history of transhumanist thought. J Evol Techno 14(3):1–30

Dekker C (2007) Stel grenzen aan het gesleutel van de mens. NRC Handelsblad 10–11 Nov 2007

Fukuyama F (2002) Our posthuman future: consequences of the biotechnology revolution. Farrar, Straus and Giroux, New York

Fukuyama F (2004) Transhumanism. Foreign Policy 144:42–43

Gavaghan C (2007) Defending the genetic supermarket: law and ethics of selecting the next generation. Routledge-Cavendish, New York

Gesang B (2007) Die Perfektionierung des Menschen. Walter de Gruyter, Berlin

Glover J (1984) What sort of people should there be? Pelican, London

Harris J (2007) Enhancing evolution: the ethical case for making better people. Princeton University Press, Princeton

Kass L (2002) Life, liberty, and the defense of dignity: the challenge for bioethics. Encounter Book, San Francisco

Kurzweil R (2005) The singularity is near: when humans transcend biology. Viking adult, New York

Kurzweil R (2006) Reinventing humanity: the future of machine-human intelligence. Futurist 40(2):39–46

Ramsey P (1970) Fabricated man: the ethics of genetic control. Yale University Press, London

Rostand J (1959) Can man be modified? Basic Books, New York. Translated by Jonathan Griffin. Original: Peut-on modifier l'homme? (Gallimard, 1956)

Sandel MJ (2007) The case against perfection: ethics in the age of genetic engineering. Harvard University Press, Cambridge

Savulescu J, Bostrom N (2009) Human enhancement. Oxford University Press, Oxford

Transhumanist Declaration (1998). http://humanityplus.org/philosophy/transhumanist-declaration/

Warwick K (2003) Cyborg morals, cyborg values, cyborg ethics. Ethics Inf Technol 5:131–137

Chapter 2
Historical and Philosophical Reflections on Natural, Enhanced and Artificial Men and Women

Christoph H. Lüthy

Abstract This chapter considers human engineering from a historical and philosophical perspective. Engineering suggests artificiality and thereby takes us to the issue of 'nature versus nurture'. Must any intervention in natural growth and development patterns be considered 'artificial'? Humans belong to a domesticated species, and the notion that human beings are shaped through both their biological heritage and their upbringing is as old as Western thought itself. Ideas about the manufacturing of humans—homunculi, golems or Frankensteins—remained usually in the sphere of pure speculation. Only in the twentieth century was the old suggestion, first formulated by Plato, that it would be profitable to breed humans like cattle first translated into political measures, as a consequence of social Darwinist ideas. Historically, we find ourselves in a unique position because we are, for the first time, able to change the human body through technological means. While many current practices can still be defined as therapeutic interventions, as genetics and artificial intelligence are further developed, the ethical issues involved in their application will inevitably become more complex. It is of great importance that before science and technology present us with unpleasant choices, society itself, as well as legislators and scientists, should determine where to draw the line between desirable and undesirable modifications of human nature.

As the Introduction to this volume has indicated, current discussions concerning the perfecting, engineering, conditioning, manufacturing or enhancing of humans

---

Translated by Samuel van Kiel.

C.H. Lüthy (✉)
Faculty FTR, Center for the History of Philosophy and Science,
Radboud University Nijmegen, Erasmusplein 1, 6525 HT, Nijmegen, The Netherlands
e-mail: c.luethy@ftr.ru.nl
URL: http://www.ru.nl/philosophy/chps/current_members/luthy/

mix facts and fictions and tend to view actual scientific and medical practices in the light of possible and imaginary future developments. For this very reason, the debate often suffers from a lack of conceptual clarity.

Let us therefore begin by unpacking some of the underlying concepts. The most important idea that calls out for analysis is that of the un-tampered with, 'natural' man, the presumed subject or victim of all technical interventions. The notion of 'artificial man' implies, after all, that there is such a thing as a 'natural man', from whom he can be distinguished.

Traditional Ways of 'Making Humans'

To begin with, let us recall that traditionally, Christianity considers man to have been made in a non-natural manner—namely by God. According to the biblical descriptions in Genesis 1 and 2, the Creator 'made' Adam and Eve (*fecit*, in the Latin of the Vulgate), together with the world and its other inhabitants. Conspicuously, the notion of nature and man as products of the divine Artificer, the *summus Artifex*, seems to suggest that humans, even when they dwelt in the most natural of places, Paradise, were artefacts. To disentangle the apparent contradiction inherent in the notion of a natural artefact, theological jargon distinguishes human acts of 'making' (from existing materials) from divine acts of 'creating' (out of nothing, *ex nihilo*).

It is worth keeping in mind the distinction between fabrication and creation when considering the claim made by religious critics that today's geneticists are 'playing God'. Strictly speaking, this claim may be rejected by pointing out that geneticists are unlike God precisely because they cannot 'create out of nothing'; at best, they intervene in, and modify, existing materials. Their ways of making would, therefore, differ fundamentally from the Creator's.

However, when the accusation of playing God is levelled against genetic engineers, this does not refer to the act of creating as such, but rather to the alleged modification of the *essence* of a God-given human nature. Nick Bostrom summarises the logic of the charges as follows: 'playing God, messing with nature, tampering with our human essence, or displaying punishable hubris'. But note that this concatenation of accusations is hardly self-evident. To begin with, the belief in static, species-related essences is not Judeo-Christian, but Aristotelian in origin, and it is doubtful whether a modern theologian needs to subscribe to it. Even present-day supporters of 'intelligent design' are usually content with a God who created natural species in such a way that they may continuously change from within, and in so doing propel evolution in the process. More specifically, as far as the question of the essential nature of humans is concerned, both Aristotelian philosophers and Christian theologians would concur that it is to be found in the soul (which to the Aristotelian represents the specific form of man)—yet, no one accuses geneticists of tampering with the soul. Lastly, it is also to be doubted whether such an essentialist view on natural species should be upheld, even from a theological standpoint, in a period in which most educated contemporaries conceive current species to be the result of an evolutionary process. In short, associating the modification of human bodies with playing God is dubious from a theological point of view and outdated philosophically.

When we worry about contemporary or future attempts to make artificial humans, we obviously do not intend to refer to divine acts. We also exclude the other obvious possibility of making humans; namely, parents' traditional way of 'making babies' through copulation. Incidentally, the seventeenth-century Flemish philosopher Arnold Geulincx correctly pointed out that the expression 'making babies' is odd and linguistically inaccurate. A potter who 'makes a pot' has acquired the skill for his trade and is able to 'make' a pot precisely because he knows *how* to make one. A man and a woman, on the other hand, have on the whole little understanding of what happens between that enjoyable night and that moment, roughly 9 months later, when a brand new, tiny but complete human person materialises from the woman's womb. Certainly, the parents have not 'made' the child in the common sense of the word.

Having now excluded divine acts of creation and human acts of procreation, let us examine the types of 'making humans' that are suggested in current debates. What we find in all discussions are the following two elements: (1) human action (as opposed to divine intervention) and (2) artificial methods of production (as opposed to natural ones).

While these definitional restrictions may prove useful, they also raise new questions. Specifically, it remains to be seen whether 'artificial' and 'natural' are really opposites, as one would expect. When organic shops conjure up the notion of 'natural foods', they obviously do not intend the opposite of 'artificial'; in fact, the organic cabbage that is sold is not the opposite of non-organic cabbage. Rather, what the shop owner means is that his or her cabbage has not been sprayed with chemicals. Applying this reasoning to human beings, it is clear that someone is called 'natural', instead of 'artificial', when he or she has not been 'treated'. There is, however, a problem with both the cabbage and the human being: neither life form grows in the wild—both are cultivated products! The question of distinguishing between treated and untreated, and between natural and artificial, in human beings, takes us to the well-known debate about how much of our behaviour results from nature and how much from nurture. While this age-old debate needs not to be examined here, it must be obvious that we humans have never been entirely 'untreated' and 'organic'. We are, perforce, socialised creatures who are raised in a cultured, and therefore unnatural, environment.

But if the definition of a natural human being remains elusive, then the same must needs apply to the definition of an artificial human being! Indeed, in the literature, we encounter a surprisingly diverse range of examples for the term 'artificial human', of which the most important are the following:

- someone fertilised in vitro;
- someone with different ('modified') DNA with respect to that of his or her parents;
- someone cloned from the cells of another individual;

- someone who is 'enhanced' in a number of ways, through genetics or techno-logical implants, in order to perfect his or her looks and abilities;
- someone composed of organic material and a neurocomputer that controls cerebral functions;
- someone who is not made up of any organic materials and instead is a machine that simulates human behaviour perfectly.

These six examples have little in common. The first five display a progressive gradation in modifying human material; with the last example, modification has made way for an entirely artificial copy.

It is clear from what has been discussed so far that we use the term 'artificial' as an umbrella term that applies to any kind of intervention in natural patterns of growth and development. Is it legitimate, however, to equate artificiality with any type of intervention? Is such an equation credible in our modern society, which, all the way from prenatal diagnostics to the nursing home, protects and prolongs life by technological means? In other words, does life in modern society, from 'good clothes, a well-stocked larder, a TV set, a car, a house and so on, all within the existing order' (Herbert Marcuse), take place within an essentially unnatural, indeed artificial, context? Or, put even more radically, is the cunning *homo sapiens*, who is 'knowing' by definition and therefore a dexterous tool creator, not always actively helping, improving or denying the natural through ways of the artificial?

Throughout the centuries, people have without any hesitation accepted that human beings have been shaped, conditioned and improved through natural selection, education and indoctrination. This acceptance, then, takes us to the following question: what is it that is substantially different about contemporary attempts at improving human beings?

This question becomes particularly difficult to answer when we consider con-temporary humanity as the merely provisional and transient result of the evolu-tionary power of selection. If *homo sapiens* is itself the result of a process of selection which has continuously preferred individuals and species made up of favourable attributes, then to what extent are modern attempts at improving human beings anything else than a conscious execution of 'natural' forces? We shall have to return to this question below.

Breeding Humans

The way in which the current debate talks about engineered or artificial humans suggests that science is tampering with the 'natural form' of human beings. Yet, we have just questioned whether one can presuppose that such a thing as a 'nat-ural' human being exists. Concerning plants and animals, we can distinguish 'wild' species from 'domesticated' ones: the first reproduce freely and without restraint, while the second are cultivated or bred. We humans would seem to belong to the second kind. Humans do not grow up in the wild, such literary examples as Mowgli or Romulus and Remus excluded. *Homo sapiens* is an 'eminently domesticated animal', to invoke an expression once used by Charles Darwin.

Farmers have known from time immemorial that they, using techniques such as grafting (for plants) and breeding (for animals), can improve the quality of new stock by combining the parents' desired attributes. The same techniques can obviously be applied to the human species. The most famous proposal to apply breeding techniques to human beings is found in Plato's *Republic*, written in the fourth century BC. Plato suggests that it would be advantageous for a state if the ruling classes were produced using the same criterion that farmers use in improving their animal livestock. Men and women with the best physical and mental attributes should be selected and encouraged to reproduce—outside of any family context, remarkably enough—and their progeny would subsequently be trained to become the ideal members of the ruling class. Aware of the difference between humans and livestock, however, Plato developed his scheme to go beyond breeding alone. He accurately described the intellectual, athletic and psychological programmes through which children would have to pass on their way to perfection.

The notion of the profile of an individual human as being as much determined by inherited and psychological features as by his or her physical, emotional and intellectual characteristics is therefore clearly as old as Western philosophical and scientific thought itself. The expressions 'well-bred' and 'of good extraction', used to typify a person conforming to acclaimed standards of behaviour, derive from this original farmers' experience, which had already risen to Platonic heights more than two millennia ago.

Plato's project was not carried out at the time, and it is somewhat surprising to historians to find that the concept of breeding humans as one would breed horses and cows was ignored even in times when Plato's philosophy was *en vogue*. In fact, even the historians' surprise has historical precedents. In Tommaso Campanella's *The City of the Sun* (1602), we encounter a senior official of a utopian state 'who takes care of generation, and of the union of males and females in such a way that they produce a good race. And they laugh of us because we look after the race of dogs and horses, but neglect our own'. Only after 1859, thanks to the publication of Charles Darwin's *The Origin of Species*, did a serious debate initiate about the long-term effects of goal-oriented selection in the process of domestication, as Darwin's readers started taking the idea of breeding humans seriously. If evolution truly functioned the way Darwin described it, so some of his followers reasoned, then it would be beneficial for a society to act analogously to the way that farmers do with their livestock by guiding society's reproduction through approved directives. Imposed choice was to replace chance.

In 1865 Francis Galton exclaimed: 'What an extraordinary effect might be produced on our race, if its object was to unite in marriage those who possessed the finest and most suitable nature, mental, moral, and physical!' These extraordinary results could, however, only be obtained on the basis of a specific policy that would prevent the increase and propagation of inherited vices such as 'craving for drinking or for gambling, strong sexual passion, a proclivity to pauperism, to crimes of violence, and to crimes of fraud'. A supposedly scientific movement emerged which called itself 'eugenics' (literally 'well-born'), which was influenced by voices like Dalton's; voices which were grouped together under the term 'social Darwinism'. This movement worked towards 'the self-direction of human evolution'.

In the early decades of the twentieth century, scientific and political leaders in various countries propagated measures to 'self-direct human evolution', according to the slogan of the eugenics movement. Initially the movement advocated policies, to be implemented by the government, of facilitating civilians who had the desired characteristics (positive eugenics) and of hindering the reproduction of civilians with undesirable characteristics (negative eugenics). In immigration countries such as the United States, this meant for instance that certain values were attributed to specific immigrant groups. A policy was adopted which encouraged the influx of families of highly valued races and discouraged an influx of any of the other races. Charles Davenport, director of The Station for Experimental Evolution in Cold Spring Harbor as well as of the Eugenics Record Office, wrote as early as 1910 that 'society must protect itself, as it claims the right to deprive the murderer of his life, so also it may annihilate the hideous serpent of helplessly vicious protoplasm'.

Notoriously enough, Nazi Germany took this concept further than any other nation: to them racial cleansing justified the ruthless extermination of so-called inferior races, a programme that complemented the positive strengthening of the Nordic races through institutions of reproduction such as *Lebensborn*. While eugenics has thus in its worst manifestation led to genocide in the name of racial cleansing, it has more frequently implied sterilisation programmes for individuals with a mental handicap, which were implemented not only in communist countries, but also in Sweden and elsewhere. Moreover, even after World War II, a series of European countries ran programmes that systematically snatched away gipsy children, who could subsequently be adopted by 'regular citizens'.

Such measures are without a doubt attempts at steering the makeup of society through governmentally implemented choices in reproduction. It is, however, unclear whether the results may be called 'artificial'. Eugenicists themselves argued that their measures only reinforced, or gave direction to, a natural process of selection that is, unconsciously in nature and consciously in human societies, omnipresent. Did not nobles traditionally marry other nobles and landowners other landowners, while the affluent could afford to marry the beautiful and healthy of lower extraction, thereby enhancing the vigour and looks of their own families? And did not, by contrast, those who were redundant and physically or mentally less endowed die of hunger? Were not the superfluous sent to die in battles as mercenaries or confined to monasteries where they were deprived of the possibility of reproducing themselves?

There is certainly some truth to the claim made by the proponents of the eugenics movement that they did not propose anything 'unnatural'. They merely claimed to carry out the work that 'nature unrestrained' would have done if left to its own device; namely, exterminating the weak and destroying those who in natural surroundings would not be fit to survive. In fact, the movement claimed it was rather the behaviour of civilised society that ran counter to the demands of nature, by setting up orphanages, poorhouses, psychiatric institutions and hospitals, which only served to keep alive those who were in truth unfit to live. This type of reasoning, which is often traced back to the writings of Herbert Spencer, asserts that the brutal measures suggested to government merely represent a return to natural law and counterbalance the artificial and moreover detrimental forces of civilised behaviour.

Whoever intends to reject such social-Darwinist reasoning—and to reject it has since the atrocities of World War II of course become the universal norm—will have to do so on the basis of ethical reasoning. Ethical rules, however, are not rules concerning our natural and ordinary behaviour; rather, they concern how we *should* behave. The nineteenth-century Darwinist Thomas Huxley was quite right about this point: ethical norms are the absolute opposite of 'naturalness'. If one accepts the supposedly natural law of the survival of the fittest as the 'natural' touchstone for behaviour, then one has to accept a 'gladiatorial theory of existence', in which the strong have the right to kill the weak. Ethics is, in this view, an antidote; a collection of rules that allow us to defy nature's ruthlessness.

What does this short historical survey teach us? We may assert, it would seem, that initially, positive eugenics simply took Plato's old suggestion seriously: the only way to produce better humans was by crossbreeding suitable individuals. The twentieth century offered new, previously unimaginable opportunities through scientific and technological advancement. Literature had, as so often happens, already foretold these opportunities. In 1932, even before Hitler, with all his obsessions with racial purity, had ascended to power, but in the heyday of eugenics, Aldous Huxley published his famous *Brave New World*, which to this day continues to appeal to our scientific imagination. It is routinely overlooked that the core of Huxley's novel is once again Plato's proposal to breed social classes from within an institutionalised governmental apparatus and outside of familial structures—although reproduction in Huxley's world takes place in vitro, and not, as in Plato's *Republic*, in vivo. What renders these nightmares relevant is, however, that they wed Platonic fantasies with twentieth-century science and technology. Indeed, they sketch an insidious mix of biological reproduction methods and psychological techniques of imprinting, which are aimed at conditioning both individual and group behaviour. To these techniques, Huxley adds the happiness-inducing drug 'soma', which is applied whenever conditioning alone proves inadequate to achieve the state-imposed individual feeling of happiness.

Conditioning People

Huxley's combination of biological, psychological and drug-related methods of conditioning leads to different kinds of perfecting or engineering human behaviour. As mentioned earlier, the eugenicist Charles Davenport believed that in order to improve society, people needed 'to annihilate the hideous serpent of helplessly vicious protoplasm'. His claim that our genetic makeup fully determines who we are marks one of the extremes in the nature-versus-nurture debate. It is astonishing to realise that the other extreme was voiced in the same time period and in the same country, by the American behaviourist John B. Watson (1930), p. 104:

> Give me a dozen healthy infants, well-formed, and my own specific world to bring them up in and I'll guarantee to take any one at random and train him to become any type of specialist I might select—doctor, lawyer, artist, merchant-chief, and, yes, even beggar-man and thief, regardless of his talents, penchants, tendencies, abilities, vocations, and race of his ancestors.

Watson completely contradicts Davenport, giving a diametrically opposite interpretation to what it means for a human to be conditioned. In fact, he is much closer to traditional views of what it means to perfect or indeed 'make' humans. Throughout history, the malleability of human nature has been mostly associated with the ability to shape speechless infants through upbringing into well-functioning, morally upright and responsible adults. By analogy, even today, we call a 'self-made' man or woman someone who is the architect of his or her own fortune. Irrespective of how we regard this malleability of the human character, which allows particularly the young to develop their personal character in response to upbringing and to pedagogical, religious and ideological indoctrination, it represents without a doubt the oldest form of modification. *Homo sapiens* has ever since it first appeared as a species lived up to its own name by passing on techniques, knowledge and convictions through education. Raising the young has always involved using carrots and sticks—encouragement and punishment. In his *Philosophical Investigations*, the philosopher Ludwig Wittgenstein compares the acquisition of a language, the first step in human education, to the training of a dog: in order to understand a language one must first of all develop the correct behavioural response to hearing a word.

Now, in the context of our present analysis of 'natural' and 'artificial' human characteristics, it is important to realise that in many cultures education is seen as an act of suppression of natural tendencies. In a Christian context, for example, it has long been common to think that the body possesses a natural tendency towards sin. This tendency was the inherited consequence of the fall of man, namely 'original sin'. Within this context, education and self-taught spiritual practice up to and including the act of self-flagellation were considered the appropriate remedies to suppress natural but sinful bodily urges and to elevate the mind above them.

At the same time that educational conditioning was being taken out of its traditional religious context, it became a central question during the transformation of psychology into an experimental science. Pierre-Jean Cabanis writes in one of his *Rapports du Physique et du moral de l'homme* in 1802, vol. III, p. 433:

> Without doubt, it is possible, by a plan of life, wisely conceived and faithfully followed, to alter the very habits of our constitution to an appreciable degree. It is thus possible to improve the particular nature of each individual; and this goal, so worthy of the attention of moralists and philanthropists, requires that all the discoveries of the physiologist and

2 Historical and Philosophical Reflections on Natural 19

physician be considered. But if we are able usefully to modify each temperament, one at a time, then we can influence, extensively and profoundly, the character of the species, and can produce an effect, systematically and continuously, on succeeding generations.

Rising ideologies brought with them attempts at 'an effect, systematically and continuously, on succeeding generations', through collective education. Such attempts were seen as a way of producing the kind of citizen that a society wanted. An extreme example of such collective, educative conditioning is the Soviet Union, which rejected the blind evolution described by Darwin and preferred a crude version of Lamarckian inheritance (known as Lysenkoism in the period 1930–1950), as efforts were underway of producing a *homo sovieticus* mostly through educational means, and only in the second instance through biological selection.

It is, however, useful to remember that the notion of collective, educational conditioning is also present in democratic societies today. Cornips and van Asselt, in their contribution to this book (Chap. 4), show that policy makers in the Netherlands quite generally subscribe to the notion that the steering and modification of human behaviour can be achieved through laws, economy and education.

In sum, then, separating naturalness from artificiality is especially difficult, if not impossible, in the case of shaping and indeed producing specific types of human agents through education. It seems impossible to imagine an exemplar of *homo sapiens* that has not been denaturalised. Once again, the species name *sapiens* of the genus *homo* indicates exactly that state of knowing that separates us from a state of unreflective naturalness.

Artificial Humans from the Past

So far, we have looked at two unbroken traditions of thinking about methods of conditioning, modifying or improving humans, namely crossbreeding and upbringing. The current debate about the modification of humans does not, however, focus primarily on these well-known kinds of modification, but centres on the fear that new technologies will propel humans beyond the traditional processes of natural procreation and upbringing.

But even this fear is not a new phenomenon: artificially created humanoids have existed for some time—in the form of fiction, that is. Let us, therefore, take a quick glance at the older types of artificiality so as to pin down more exactly the place of today's spectres of artificial humanity.

One traditional type of artificial human is one that is brought to life through magic. Take, for example, the golem, a clay figure which is, according to Jewish legend, brought to life by rabbis using Cabbalistic spells. This animation was inspired by the story in Genesis 2, where God creates Adam from clay and blows life into him. No sooner is the spell removed than the golem turns to dust. This type, an imitation of creation and dependent on magic, is of course of little relevance to current expectations and fears, even though the science fiction writer Stanisław Lem did transpose the golem story to the era of robotics.

It is of more immediate interest here to examine those fake animations of statues and dolls that the ancient and early modern technicians loved so much. Long before Alan Turing came up with the 'Turing test'—are we able to distinguish a human being from a computer based on the answers given to our questions?—technicians invented all sorts of ways of tricking an audience into believing that a machine figurine was a human being. The success of craftsmen who made puppets that could imitate human gestures, in combination with certain seventeenth-century breakthroughs in the understanding of human physiology, convinced the philosopher René Descartes and his followers that even the bodies of ordinary humans were no more than mechanical devices.

Apart from the golem and the fabricated *homme machine*, history also tells us of another type of artificial human: the chemically produced human, of which the alchemists' homunculus is the earliest example. Figure 2.1 shows a Paracelsian alchemist who produces a 'little man' (homunculus). In contrast to the mechanical doll that inspired the Cartesians, it was rumoured that in order to produce a homunculus, one needed biological substances: the alchemist simulated biological reproduction by placing male seed in his flask (which represented a type of artificial womb) to set in motion a process of sublimation, using an oven to trigger the process.

Of course, no homunculus has ever emerged from the alchemist's workshop, and neither has, some centuries later, any artificially created monster escaped from any chemical laboratory: the heavily breathing creature in Mary Shelley's *Frankenstein* (1818) never left the novel's pages. Still, the homunculus, the monster from *Frankenstein* and the Faustian scientists responsible for their existence live on in our collective, literary memory as an expression of a deep-seated fear, which today may be stronger than ever: will not the results of our scientific experiments one day defy our control?

The very theme of escape and defiance is in any case a persevering literary subject. In Gustav Meyrinck's novel *Der Golem*, the clay man flees his rabbi's rule and becomes something of an immortal symbol of the wandering Jew. Similarly, Frankenstein's monster escapes its creator's laboratory. In our collective imagination, these precedents appear to illustrate how the future *Übermensch* will defy us, its makers. The title of one of Rakesh Kapoor's essays (2003) reflects this fear well: 'When Humans Outsmart Themselves'. In this essay, Kapoor describes Nick Bostrom's idea that somewhere in the next 50 years machines will be propelled by artificial intelligence—an intelligence not merely surpassing human intelligence, but capable of making decisions independently, and indeed possibly going against what we had originally intended them to do.

Exactly in the way that in previous centuries Cabbalistic magic, alchemy, chemistry, engineering and eugenics spurred literary fantasies about the genesis of artificial humans, today we encounter sundry androids, transhumans and other uncannily modified, technologically produced humans in novels and films. Have such fantasies now become more realistic than their predecessors, or are they

Fig. 2.1 A nineteenth-century engraving of Wagner the alchemist, from the second part of Goethe's *Faust*, manufacturing a homunculus

simply the mental product of our talent invariably to imagine alternative worlds and worst-case scenarios?

Each era faces the challenge of having to distinguish between realistic and unrealistic projections of current practices. This challenge has nothing to do with the run-away nature of human fantasy, but is due to the unpredictable nature of the evolution of science and technology. What makes contemporary fantasies especially hard to gauge is the fact that there are so many of them—they range from genetically manipulated humans, to brains maintained by machines, to humanoid robots—and that they traverse different, currently still separate scientific disciplines. To be sure, some of these fantasies and nightmares represent simple extrapolations of current practices. In an age in which hip replacements, organ transplants and pacemakers are normal, it is hardly difficult to extrapolate from these practices by imagining the replacement or enhancement of other body parts. Likewise, in a world where machines continue to replace skilled labourers, it is easy to predict that computers or robots will be allotted further tasks in the near future. What is, by contrast, still difficult to imagine as a real, imminent possibility is the disappearance of the boundary between humans and machines. Currently, the difference between the real pilot and the auto-pilot is still very clear, while the moment in which this difference disappears is as impossible to predict as it is to imagine.

Distinguishing Humans from Non-humans

In the context of discussions regarding the nature and definitional boundary of humans, it is once more useful to take recourse to historical precedents. Two of them are of interest to us. The first concerns the boundary between humans and animals, the second that between humans and machines. Both must be briefly mentioned here.

In the period 1450–1800, following the discovery of sub-Saharan Africa and the New World, a considerable confusion came about concerning the dividing line between humans and other organisms. Travellers' accounts of Pygmies, Hottentots and apes made Europeans wonder whether they were dealing with different types of humans, with animals or with a cross between these two realms of nature. Many commentators were reminded of satyrs and fauns, humans with a goat-like appearance as depicted on Greek vases. In Fig. 2.2, an image dating from 1763, we encounter a collection of 'anthropomorphic creatures', showing from left to right a 'troglodyte' (a cave-dweller of sorts), 'lucifer' (a type of devil), 'satyr' and 'pygmy'. Were these forms all human? Where did one have to draw the boundary between human and non-human? In fact, when Europeans first encountered the orang-utan and learned from the locals that this name means 'forest man', they classified him accordingly. In Fig. 2.3, one sees a male courting a female, in a particularly Dutch way, namely with a tulip. The legend explains that what we see is a 'forest satyr called orang-utan'.

When in the eighteenth century Carl Linnaeus drafted his famous *Systema Naturae*, he classified each species into (1) kingdom; (2) class; (3) order; (4) genus; and (5) species. Humans, he decided, belonged to (1) the animal kingdom; (2) the vertebrate class; (3) order of primates; and (4) to the genus *homo*. But when having to decide on the species name, he faced a dilemma. In the early editions of his *Systema*, he split the genus *homo* into two kinds which he called 'diurnal' and 'nocturnal', classifying most species as active during the day, and others, like the orang-utan, as active at night. Importantly, Linnaeus indicated that on anatomical grounds, he had no reason to place humans into a different genus than orang-utans. Later editions show a remarkable shift: Linnaeus split the genus *homo* into two main species, *homo sapiens* and *homo monstrosus*. Particularly striking is the fact that the orang-utan is classified as *homo sapiens* (though belonging to a sub-species of 'wild human'), while mountain dwellers, Hottentots, Chinese and Indians are classified as 'monstrous man'. In other words, in Linnaeus' eyes some human races were less human than certain primates.

Fig. 2.2 'Anthropomorphic Creatures'. From: Carl Linnaeus, 'Anthropomorpha', in idem, *Amœnitates Academicae* 6, Stockholm, 1763, p. 76

Fig. 2.3 Forest satyrs. From: Peter van der Aa, *Icones arborum, fructorum et herbarum exoticarum*, Leiden, 1700, plate 77

From our point of view, all of this is obviously somewhat of a laughing matter. Yet, Linnaeus' confusion is brought into a new perspective when taking into consideration contemporary geneticists, who claim that there is little difference between humans and chimpanzees genetically, and behavioural sociologists, who are experimenting in teaching primates to use tools and language.

Moreover, in science fiction this eighteenth-century confusion has recently witnessed a strong revival. Take *Star Wars*, for example, where normal-looking humans work together with intelligent primates and robots. Most of the time it is intuitively clear whether or not someone belongs to the species *homo sapiens*. In other cases, which are naturally the more interesting ones, this distinction is unclear, and such cases go to show just how unstable dividing lines really are. Modern robotics examines these divides, as Catholijn Jonker and Annemiek Nelis show in Chap. 7—divides, moreover, that may well have important legal consequences, as Bert-Jaap Koops proves in Chap. 12.

It is particularly interesting in the present context to examine the early modern controversies about the dividing lines between humans and non-humans. First, because these debates show that the status and boundaries concerning our species have been unclear even before our technologically advanced age. Second, because a fundamental touchstone was used to distinguish human from non-human; namely, the immortal human soul, which was taken to constitute a uniquely human trait. Despite the fact that this soul could neither be seen nor measured, and thus could not be tested empirically, it did provide a unique ontological criterion for distinguishing humans from animals.

The immortal soul is today no longer considered the central criterion; self-consciousness has replaced it. This new benchmark, however, is far more problematic than the immortal soul. Not only is there a consensus among experts that some animals show behaviour that testifies to the presence of self-consciousness; it is also a matter of debate whether consciousness might turn out to be a necessary attribute of any highly complex neural network. Both considerations render consciousness a property that is not suitable for uniquely singling out humans.

Besides the contested borderline between humans and animals, there is also that separating humans from machines. Once again, the obvious point to start is in the early modern period. In the 1640s, René Descartes, whom we have encountered earlier, defended the notion that the bodies of animals are essentially machines that are complex enough to be self-multiplying. Inasmuch as they possess bodies that function like those of animals, humans are also just self-sufficient, multiplying machines. Yet, contrary to animal bodies, human bodies also house an immortal soul, a kind of spirit inhabiting a machine.

Descartes regarded our heart as a pump, our veins as pipes, our muscles as levering devices and our eyes as optical instruments, and he felt that it was legitimate to assume that all other parts and functions of our body could equally be explained in such mechanical ways. Was the movement of an arm towards an object anything more than a feedback device brought about by sensory input?

Descartes' view of the human body as a machine has enjoyed a strong revival in the twentieth century. For example, in his first bestseller, *The Selfish Gene*,

Richard Dawkins describes the human body as the 'survival machine' of the genes that carry its blueprint. As must be obvious, such a mechanistic view only makes things more complicated for our discussion. After all, if the human body is viewed as a machine, then it is hard to specify why a cyborg should belong to an inherently different category.

Against this view, it can be maintained that Descartes' description of man as a machine was merely intended as an analogy with an explanatory purpose. For, Descartes did not claim that humans and man-made machines were the same; he merely asserted that human beings *work* like machines while being infinitely more complicated in structure. This is an important distinction to make. The real difference between Descartes' description of humans as machines and the man-machine that some apostles of transhumanism enthuse about is this: the transhuman man–machine does not multiply naturally the way that Descartes' *homme machine* does. The transhuman, partly designed by humans, breaks away from the evolved organisms the functioning of which Descartes tried to capture with his machine analogy.

For the time being, and pending the invention of truly revolutionary types of machinery, it may be said that the dividing line between humans and robots or transhuman androids consists above all, and maybe essentially, in the inability of the latter to reproduce their own (upgraded) form. This criterion resembles apostolic succession, the Catholic Church's criterion for determining orthodoxy, whereby a church is only considered orthodox when its priests are appointed by bishops who in turn have been appointed by earlier bishops, going back uninterruptedly in a direct line to the time when Jesus appointed the apostles. Likewise, we might wish to choose as a criterion for distinguishing between 'natural' or 'genuine' and 'artificial' humans the criterion of direct descent, through natural reproduction, from historical humans (or 'palaeohumans', as Koops calls them in Chap. 12). It must, however, be clear that this definition by succession is only useful in the case of machine-enhanced humans or robotic imitations. It is, however, useless in the case of genetically modified humans, where an unbroken generational succession might be maintained despite immense modification of traits, capacities and behaviour.

The Novelty of Our Historical Situation

We have examined a range of historical case studies to help us determine in which way our own understanding of what it means to modify human nature (and all the hopes and fears attached to it) might differ from earlier notions, hopes and fears. On the basis of what has been said it may be concluded that where 'artificiality' refers to conscious crossbreeding or to the modification or conditioning of behaviour, our current expectations and apprehensions do not differ in any real sense from those of past ages. It also seems that there is little new about the dream—or nightmare—of producing humans in a laboratory. This fantasy has been around for centuries, and it has to this day remained fictional.

What does, however, distinguish today's situation from all earlier ones is the concept of *perfecting existing humans* individually or collectively through technology. In the past, people tried to create better offspring by matching parents of good stock, or people imagined androids being made in laboratories or through magic. Only in the last decades has the idea emerged of modifying existing human individuals either through biological means (by replacing malfunctioning parts or genes) or by means of electronic aids (by adding gadgets which replace malfunctioning parts, or by improving parts that already function well).

Most of the current practices fall under the overarching term 'improvement'. Various aids are already in place, from pacemakers to Viagra, from Prozac to plastic surgery. Most of these have so far been of a corrective nature. Even Marijke Helwegen, who in the Dutch newspaper *NRC Handelsblad* of 2 June 2008 is called 'the ambassador of the artificial body', insists on repair and renovation, rejecting alteration: 'I have not been reconstructed; rather, I have been renovated. I was gorgeous, and I have stayed that way'. What Helwegen suggests is that all she does is to maintain her original appearance in the face of time's pitiless rule. Other forms of improvement allow performance quality to be pushed to new heights, yet, once more, without the insertion of new characteristics. None of these practices defy traditional ideas of human nature.

The predicted linking of artificial intelligence to the brain belongs, by contrast, to a different category altogether, because if successful, it would infringe upon traditional conceptions of personal identity, which since ancient times have been associated with mental phenomena such as memory and personal convictions. As early as 1957, Oswald Wiener designed a thought experiment in his novel *Die Verbesserung von Mitteleuropa*, in which a machine brings about a smooth transition from a natural to an artificial state of consciousness. Wiener's 'bio-adapter' is a machine that is placed on the head. After measuring and registering response patterns to stimuli for a certain period, it begins to imitate them in order to adopt the tasks of the biological brain. The unison between the processes of the brain and those of the machine enables the bio-adapter to ensure that the subject never becomes aware of the slow but continual displacement of control. In time, the machine is able to reproduce all conscious states of mind.

It is a small leap from Wiener's imaginative machine that reproduces the biological brain perfectly to a machine that brings about new mental processes. Those new processes could fundamentally differ from the mental conditions we normally observe in biological brains. This idea of replacing biologically-driven functions by technologically-driven ones is what I believe to be historically new. Previous generations have given much thought to artificially created human beings, yet these humans were always different, not 'us'. They were produced in laboratories in their entirety and were not enhanced by machines.

The so-called 'transhumanists', enthusiastic as they are about the envisaged consequences of technological improvements of the current human condition, do not view this situation as particularly alarming. They are of the opinion that the human species as currently existing is not an unchangeable species to begin with, but rather a temporary step somewhere on the evolutionary ladder. They 'view human nature as a work-in-progress, a half-baked beginning that we can learn to remould in desirable ways. Current humanity need not be the endpoint of evolution', as Nick Bostrom, one of the founders of transhumanism, explains (2003, p. 493). He imagines transhumans who are not just more intelligent and learned, but also more amiable, friendly and richer in their aesthetic and introspective experiences: 'Healthier, wittier, happier people may be able to reach new levels culturally' (2003, pp. 498–499).

How does this 'technological prophecy' differ from the 'natural prophecy' by Alfred Wallace, the co-inventor of the theory of evolution, who in 1864 observed that 'the power of "natural selection", still acting on [man's] mental organisation, must ever lead to the more perfect adaptation of man's higher faculties to the conditions of surrounding nature, and to the exigencies of the social state'? The difference lies once again in the fact that fin-de-siècle techniques differ from today's. Whereas evolution used only to be 'assisted' by directing choices in reproduction, transhumanist methods propose to alter the genetic makeup of humans and, moreover, to enlarge human capacities through technological means.

It remains doubtful whether such fantasies are any more realistic than those about the alchemist's homunculus or the Cabbalist's golem. But as the German philosopher Bernward Gesang rightly emphasises in his recent study *Die Perfektionierung des Menschen*, we must be ethically prepared to handle even the most absurd situation before it presents itself. Before we are confronted with the results of science and technology, we must decide a priori and as a society in its entirety on the limits of research and its applications. Basing himself on utilitarian ethics, Gesang suggests that we allow modifications of bodily and mental enhancements within well-defined legal limits, and within a social framework, while forbidding by law any radical restructuring of the human body or of mental functions. The latter carries severe risks of leading to grave social imbalances, and are therefore undesirable on a utilitarian account. Regarding genetic manipulation, Gesang pleads for the application of methods of improvement on children as long as this does not change their human appearance, and under the pretext that empirical proof has shown such methods to have worked on adults.

Gesang urgently and convincingly pleads for the need to determine the limits of our willingness to improve or deform human nature well ahead of the advent of the technologies that might allow for the relevant practices. As has hopefully become obvious from this chapter, the ethical and political task of defining and protecting the boundaries of human nature will benefit from a look at the past and from viewing our own expectations and fears against the background of the long history of views on human perfectibility. History has shown just how liable to change ideas on the essence of human nature have been throughout the ages. At the same time, an enhanced historical awareness helps us understand where the real technological difference lies between former aspirations and current ones.

Bibliography

Bostrom N (2003) Human genetic enhancements: a transhumanist perspective. J Value Inq 37:493–506

Cabanis P-J (1823/1981) On the relations between the physical and moral aspects of man, vol 1. Johns Hopkins University Press, Baltimore

Campanella T (1981) The City of the Sun: a poetical dialogue. eBook collection EBSCOhost

Davenport C (1910) Report of the Committee on Eugenics. Am Breeders' Mag, pp 126–129

Drexler EK (1992) Nanosystems: molecular machinery, manufacturing, and computation. Wiley, New York

Galton F (1865) Hereditary talent and character. Macmillan's Mag, pp 157–166, 318–327

Gesang B (2007) Perfektionierung des Menschen. Walter de Gruyter, Berlin

Huxley T (1803/2009) Evolution and ethics. Princeton University Press, Princeton

Kapoor R (2003) When humans outsmart themselves. Futures 35:787–791

Kurzweil R (1999) The age of spiritual machines: when computers exceed human intelligence. Viking, New York

Kurzweil R (2005) The singularity is near: when humans transcend biology. Viking, New York

Lem S (1981) Golem XIV. Wydawnictwo Literackie, Cracow

Marchant J (1916) Alfred Russel Wallace: Letters and Reminiscences, vol 1. Cassell, London

Marcuse H (1965) Socialism in the developed countries. Int Socialist J, pp 150–151

Moravec H (1999) Robot: mere machine to transcendent mind. Oxford University Press, Oxford

Wallace A (1864) On the origins of human races and the antiquity of man deduced from the theory of 'Natural Selection'. Anthropol Rev, pp 158–187

Watson JB (1930) Behaviorism. W W Norton, New York

Wiener O (1969) Die Verbesserung von Mitteleuropa. Rowohlt, Reinbeck

Wittgenstein L (1953) Philosophische Untersuchungen. Blackwell, Oxford

Chapter 3
Changing the Body Through the Centuries

Theo Mulder

Abstract This chapter presents an overview of the ways in which human beings have tried to change their bodily appearance through the ages. In the Middle Ages, the body was perceived to be unclean and corrupt, because all human beings were regarded as sinful. Any preoccupation with the physical was considered as an act of vanity, as only the soul could be pure. Over the subsequent centuries, taking charge of the outer appearance slowly became more accepted, as attempts at plastic surgery were made, chiefly through nose reconstruction for syphilis patients. These kinds of efforts thus have their root in allowing people who suffered some kind of disfigurement to regain an appearance that was as close as possible to 'normal': changes were made not because people wanted to be noticeable, but average. This motivation persisted through to the first half of the twentieth century, in particular focussing on efforts to restore facial features of soldiers wounded in World War I. Recently, however, more and more individuals change their bodies due to a desire to stand out. These changes also imply questions about how far body modification should be allowed, and whether legal measures are necessary in this context.

In the novel *All Souls' Day* by Dutch author Cees Nooteboom, a sculptor named Victor examines a portrait of Queen Louisa of Prussia. He asks the protagonist, Arthur:

> Can you imagine how this woman must have walked? ... No, you can't... Clothes don't become extinct. They can be copied or saved, so we're all right on that score. But what does die out is the way people moved in those clothes. The fabric falls differently when the

---

Translated by Eveline Heij and Miranda Jorritsma.

T. Mulder (✉)
Royal Netherlands Academy of Arts and Sciences, P.O. Box 19121, 1000 GC,
Amsterdam, The Netherlands
e-mail: theo.mulder@bureau.knaw.nl

B.-J. Koops et al. (eds.), *Engineering the Human*, DOI: 10.1007/978-3-642-35096-2_3,
© Springer-Verlag Berlin Heidelberg 2013

movement is different. This woman could never have worn a bikini. She didn't have the right walk for it, it hadn't been invented yet.' 'Who did invent it?' 'Oh,' said Victor, 'time did... (p. 37)

Nothing could be more true: time destroys or changes everything. Everything we know, or think we know and everything we consider beautiful or ugly will eventually be turned upside down by time. Individuals whose bodies would have been admired in the past are now encouraged to work out and lose weight. The body is not allowed to bulge, wobble and shake anymore, but should be tight and move and obey like a machine. The human body should be beautiful, either naturally or by artificial means. It has to be brought under control in a fitness club, or given the right shape by a cosmetic surgeon. In April 2008, Dutch women who registered on the dating site Mooiemensen.com (literally: Beautifulpeople.com) had the opportunity of winning a new pair of breasts. A survey conducted in 2004 showed that 71 % of all Dutch citizens believed that by 2020 cosmetic surgery would be a conventional option for those who want to look more beautiful. Interestingly, the younger part of the population scored 78 % (Schnabel 2005). Thus, the image arises of a people who renovate and shape their bodies according to the current beauty ideal.

But why do we do this? And does this image reflect reality? These are the questions that will be examined in this chapter. The answers teach us as much—or as little—about humanity as they tell us about the time and culture we live in, as these two concepts are inseparable. Humans are the only animals that have the ability to be dissatisfied about their appearance. While the act of looking in the mirror is a very stressful experience for many of us, it does not in the least affect horses, dogs or rabbits. These animals are what they are—nothing more and nothing less. Human beings, on the other hand, are more complicated. They want to belong to a certain group or distance themselves from it; they want to stand out or blend in. Humans are not what they are, but rather what they *want* to be, are supposed to be or have transformed themselves into—sometimes literally.

The Body as the Mask of Death

This narcissistic and malleable body ideal has not always been the case. In the Middle Ages, the idea prevailed that human beings are sinners undergoing punishment. In this context, images of strong and healthy bodies were inappropriate, as was the wish to change or beautify one's body. Pope Innocent III (1161–1226) described the human race as 'food for worms which never cease to gnaw and consume: a mass of putrefaction, ever fetid and horribly dirty' (qtd. in Camporesi 1988, p. 109). The monk Bernard of Clairvaux held a similar opinion and thought of humans as 'nothing but stinking sperm, a sack of excrement and food for worms' (qtd. in Camporesi 1988, p. 78). The living body was thus seen as a mask of death. It was by definition impure, unclean and a place of rotting and decay in which the process of decomposition did not start after death but during life. Not

3 Changing the Body Through the Centuries

only was it filled with slimy organs, bodily juices and faeces, it was also crawling with maggots and worms. The body smelled, and this scent was associated with rotting (Labrie 2001, p. 76). Women were often blamed: they were, after all, the gateway of the devil and the cause of the downfall of God's image—man. The revulsion against the human body was mainly directed at sexuality, that mysterious and omnipresent force behind human behaviour that can excite both lust and fear.

This cesspit of vice that was the human body could only be controlled through chastisement. Flagellants—monks who scourged themselves with whips—could be found travelling through European cities. The plague and leprosy claimed hundreds of thousands of victims, thus emphasising the defencelessness and worthlessness of humans and their bodies. Lepers were not only kept outside the city gates, but also outside society. They were pariahs, contaminated by the venom of the infernal serpent. In the Bible, they were symbols of evil, ruin, death and downfall. Heretics belonged to the same category. They were the murderers of souls, and along with arsonists, murderers and homosexuals they deserved the ultimate punishment: burning at the stake.

It was a time of great fear: fear of the unknown, the foreign and diseases—which were naturally, and invariably, transmitted by foreigners. As a consequence, humans developed a longing for the clean and pure. Not in the sense of hygiene—it would still be several centuries before 'purity' would come to imply clean water and the washing and grooming of the body—but purity of belief. Only the 'pure in heart' (Matt. 5.8) would be allowed to behold the purity of God. The doctrine of the pure heart thus forms the core of Christianity. Purity of heart automatically implies a strong dislike of the body, which explains why taking care of that body was considered as a blasphemous vanity. Saint Peter Damian gives the example of a woman who paid too much attention to her appearance and was gruesomely punished for this. One day God took away her beauty, which only served as evidence of her mental corruption, and changed her body into a putrefying mass spreading an unbearable stench. The Bible, and in particular the Old Testament, was considerably milder in its treatment of the human body than were those who based their authority on it.

The bleak image of the Middle Ages presented here should be taken with a pinch of salt. As is the case in every era and society, there was no cultural uniformity that extended to all layers of society. Medieval society was rigidly class-based, but this is not unique to the Middle Ages. While the clergy used its power to influence the people through an ideology of guilt and fear, court culture was preoccupied with colour, detail and beauty. A good example of this can be found in the work of the Limbourg brothers, three famous Dutch manuscript illuminators. Umberto Eco also disagrees with the conception of the Middle Ages as the 'dark ages'.

However, it is important to keep in mind that the awe of beauty that was invoked by art was immediately translated into a sense of religious fulfilment and closeness to God. In his book *The Autumn of the Middle Ages*, famous Dutch historian Johan Huizinga mentions that *true* beauty can only be assigned to God; the world and its creatures can be *venustus* (nice, pretty) at best. Huizinga also emphasises the strong disapproval of frivolities. When polyphonic music was first introduced in church, many leading clergymen objected to it as being 'comparable to curled hair on a man or to pleated garments on a woman; sheer vanity' (Huizinga 1919/1996, p. 322).

Nevertheless, there still existed a body-oriented clothing culture. In court culture, garments were decorated with precious stones and bells or coins. During Louis XI's entry into Paris in 1461, the Charolais horses had many large bells attached to their horsecloths. A duke of Cleves even received his nickname *Johenneken mit den Bellen* (John with the bells) after he returned home from the court of Burgundy dressed according to this fashion.

Besides the melancholy of the courtly love poets, there was also a literature which depicted folk life by means of farce and other types of crude comedy and which by no means shunned the physical.

The late Middle Ages showed a singular contradiction between a strong sense of shame and, at the same time, an astonishing leniency towards the body and the bodily. During Philip the Good's entry into Ghent in 1457, there were 'Sirens', entirely naked and with loose hair, swimming in the river Lys to greet him. Jean de Roye describes a similar spectacle during Louis XI's entry into Paris in 1461: 'And there were also three very handsome girls, representing quite naked sirens, and one saw their beautiful erected, separate, round and hard breasts, which was a very pleasant sight, and they recited little motets and bergerettes' (Huizinga 1919/1996, p. 374).

Individual Autonomy

During the fifteenth century a change took place. Not only was there a revival in the production of images of the human body outside the context of religion and suffering, but enlightened citizens also considered it their right to change their bodies according to their own ideas. Giovanni Pico della Mirandola (1463–1494) emphasised the importance of individual autonomy. He considered illnesses and deformities not as punishments from God, but as twists and turns of fate which could be influenced. In his famous oration *On the Dignity of Man*, Pico has God tell Adam:

> Neither heavenly nor earthly, neither mortal nor immortal have We made thee. Thou, like a judge appointed for being honorable, are the molder and maker of thyself; thou mayest sculpt thyself into whatever shape thou dost prefer. Thou canst grow downward into the lower natures which are brutes. Thou canst again grow upward from thy soul's reason into the higher natures which are divine. (p. 5)

In his book *Making the Body Beautiful: A Cultural History of Aesthetic Surgery*, Sander L. Gilman describes the importance of this humanist point of view. Individuals partially regained the right to make decisions about their own bodies.

While Pico della Mirandola's ideas were radically humanist, it would take more than a century before they were put into practice, in a very literal manner, in the reconstruction of noses. This was done by Gaspare Tagliacozzi (1545–1599), who

3 Changing the Body Through the Centuries

is considered to be one of the 'founding fathers' of plastic and cosmetic surgery in Europe.

In the century between Pico della Mirandola and Tagliacozzi, a great syphilis epidemic raged in Europe, causing many patients to lose their noses and leaving them in want of a replacement. Tagliacozzi introduced a revolutionary surgical technique, using a skin flap from the upper arm to reconstruct the nose. This was a hazardous affair, however. For one thing, the procedure was performed without anaesthesia in a time without any knowledge of infections. In addition, the new nose was not very reliable: in cold weather it turned a lead grey colour, and when the owner blew his nose too hard, he risked blowing it off and becoming noseless once more.

Why was the nose considered so important? Disfigurement was not uncommon in medieval cities, and it could easily be ignored. A decaying nose, however, was the most feared—for the most revealing—of all symptoms of syphilis, a disease which originated in America. The absence of the nose clearly revealed a person's flawed character, as their immoral past was literally written on their face.

Slowly, this aversion spread to include all noses that did not fit the ideal. The perfect nose had a nose ridge that made an angle of a hundred degrees with an imaginary horizontal line. The Dutch anatomist Petrus Camper (1722–1789) was one of the first to develop the idea that a small, flat nose was a sign of inferior race. Camper was not the only one interested in the relationship between face and character, however. He wrote during the rise of physiognomy, a science based on the assumption that a person's outer appearance is an accurate reflection of their character or personality. Camper constructed a racial theory based on the form of the nose, introducing the so-called 'nasal index' which measured the nose's angle relative to the face. This index was used by many contemporaries, including his son-in-law Theodor Soemmering, to create racial hierarchies. It is hardly surprising that the African race, together with the primate, was placed at the bottom of this scale, while the white European Greeks were situated at the top.

Soon, the Jewish nose joined the African one at the bottom of the hierarchy. The nose became an abstract sign for race and heritage, and both Jewish and African people wore their heritage in their faces. Jews were witches, devils. Not only did they possess a recognisable nose, but they also had the evil eye, and at night they could transform into animals, spreading a beastly stench.

The size and form of the nose became pawns in a heated moral debate. Surprisingly, this extreme attribution of meaning was not short-lived, as the nose continued to be a topic of debate for over 200 years.

In the sixteenth century, the nose became an essential organ due to the fact that it cannot deceive its owner. It is through the nose that smells reach us, enabling us to distinguish the pure from the impure. It thus became an important aid in the distinction between social classes. While the upper classes and one's own social group were characterised by a pleasant aroma, the poor and the foreign were associated with foul odours. The seventeenth-century physician Paulini cleverly used scent to support his ideas on class society. He gives the example of a skinner who fainted while smelling the overwhelming fragrances of an apothecary, but woke up again when smelling his pelts. This was all the evidence Paulini needed: the nose never lies and provides the ultimate proof that social class is a natural phenomenon and should not be questioned (see also Corbin).

Jacques Joseph

Jacques Joseph (1865–1934) was born as Jakob Lewin Joseph, but decided to change his Jewish name when he began to study medicine in Berlin. Joseph was an orthopaedic surgeon and worked under one of the best surgeons of the time: Julius Wolff (1836–1902). Joseph was not interested in conventional orthopaedics, however, but rather in the possibility to change the human body by means of medical intervention. When he performed an operation on a Jewish child with large, protruding ears in order to place the child's ears closer to his head, Wolff fired him. According to Wolff this was not conventional surgery, but cosmetic surgery. Joseph had used his surgical knowledge for the wrong purpose: vanity. The fact that the child was constantly teased because of his Jewish 'Moritz' ears was considered irrelevant in this respect.

In Berlin Jacques Joseph started his own clinic, and he became the founder of present-day rhinoplasty (commonly known as 'nose job'). He changed so many noses and was so successful that the inhabitants of Berlin started to call him *Nasen-Joseph* ('Joseph of the noses'). After a while he also started to operate on ears and other human body parts, allowing their owners to blend in with the crowd. For this was what most of his patients wanted: to no longer stand out as Jewish and become a part of the anonymous, unharassed urban mass. Jacques Joseph provided this possibility. This form of surgery, therefore, had nothing to do with beauty ideals but with the desire not to stand out, or at least not to look Jewish and as a consequence suffer insults and persecution. Here again we encounter the importance of noses. Joseph's patients wanted nose jobs, because the shape of the nose was charged with normative significance—theirs were Jewish noses. And while Jacques Joseph changed the noses of the Jewish population in Berlin, John Roe changed the 'pug noses' of the Irish in New York.

It was not only the nose which was considered as a giveaway organ; ears were equally suspect. Ears are not just visible organs that can detect sounds, but they have also been the subject of many racial (and racist) theories. At the end of the nineteenth century, anthropologist Hans Günther wrote that Jews have big, protruding ears with fleshy lobes: the 'Moritz' ears mentioned above. In Heinrich Mann's novel *The Loyal Subject* (*Der Untertan;* 1918), Jadassohn finds his appearance too Jewish because of his 'huge, red, prominent ears' (p. 86). He goes to Paris to have his ears 'de-jewified'—or, in other words, to have their size reduced.

Jacques Joseph's existence took place on the fringes of the surgical establishment. This changed during World War I, when he was able to deploy his knowledge of face-changing surgery to help the thousands of soldiers who returned from the trenches with ruined faces. Their wounds were terrible. Major parts of faces had been shot away; pieces of skulls, jaws, eyes, noses and cheeks had disappeared. These soldiers were not dead; they were still alive and they were young. After every battle, thousands of victims were brought to hospitals on all sides. After the battle of the Somme in 1916, 2,000 gruesomely maimed soldiers were brought to the English surgical centre for facial and jaw surgery in France, of which dentist Varaztad Kazanjian was in charge. The same was true for all parties that fought at the front. In Germany, Jacques Joseph's reputation rose to unprecedented heights and after World War I his status was unrivalled.

The Strive for Inconspicuousness

In the early twentieth century, cosmetic surgery was not so much concerned with the enhancement of beauty as with the alteration of physical appearance. Until World War I, it focused mainly on the face and the reduction of notable and undesirable ethnic facial features.

The rise of cosmetic surgery complemented the late-nineteenth-century biological theories on adjustment or elimination, change or exclusion remarkably well. This was exactly what cosmetic surgery made possible: adjustment to 'the normal' and removal of deviations from the norm. Happiness depended on acceptance by the social group in which a person lived. Surgeons removed a shape's sharp edges in order to make it blend in more with the other shapes. After all, average shapes are never discriminated against; only someone who stands out attracts public judgement. The same applied to war victims: they had to be changed in order not to attract attention or cause fear and horror, so that their families would not have to confine them to the house.

This image is not only applicable to the early twentieth century but also to later periods. After World War II, a market for the westernisation of eyelids established itself in Asia which is still thriving today. Out of the 2.8 million surgical procedures conducted annually in the US, almost 20 % is performed on non-Western clients. Many of these interventions constitute a form of ethnic surgery: the removal of prominent ethnic features in 'exchange' for features of the dominant racial culture.

This process is similar to that of Jacques Joseph in early-twentieth-century Berlin. In his book, Gilman uses the term 'passing': surgeons have to make sure that someone can successfully pass for a member of the dominant culture. Kathy Davis provides several examples of women who wish to vanish into inconspicuousness. Conversations with Dutch women who underwent cosmetic surgery support the notion that such procedures are not motivated by a desire to attain beauty or meet a certain beauty ideal. These women did not want to be more beautiful, or different; they wanted not to be different, to fit in.

Human beings have the natural inclination to mirror each other's behaviour. This behaviour is not acquired, as neonates are already capable of mimicking their parents' facial expressions (Meltzoff and Moore 1983; Mulder 2007). Nor is it an age-related effect, for such automatic copying behaviour can be noticed at every age. It is a common phenomenon during meetings; if you start paying attention to it, you will notice how often your behaviour is mimicked by your conversation partner. This includes not only movements, but also facial expressions, intonation, speech rate and even respiration. It is possible to suppress this kind of behaviour once you are aware of it, but it often remains unnoticed. Humans want to look alike.

The imitation of other people's behaviour occurs more often when the other person stands higher in the social hierarchy or is well-liked. The social function of imitation becomes even clearer when someone is placed outside of their own familiar group and needs to be accepted by a new group. In such circumstances, individuals will display more imitation behaviour because mimicking will help in their acceptance into the new group. Imitation of both behaviour and appearance is an important aspect of social bonding: imitation increases resemblance and resemblance heightens sympathy (Decety and Batson 2007). Strategically, merging with the other, the desire for the average, is not a bad choice. For example, research shows that individuals with average faces are considered to be most attractive. Humans do not look for extremes (Thornhill and Gangestad 1999; Valentine et al. 2004).

Sometimes similarity is enforced. Gilman gives the example of 1950s America, where individuals whose appearance deviated from what was considered the norm, for instance due to war injuries, were forced to adjust their appearance. Medical interventions—pharmaceutical or surgical—were used to change their bodies in order to make them look more normal. A later example is the Chicago municipal code of 1966 (code 36–34), which prohibited individuals who were 'diseased, maimed, mutilated, or in any way deformed so as to be an unsightly or disgusting object' from appearing in public, on penalty of fine (Gilman 2000, p. 24). The code, initially introduced to prevent freak shows, was not abolished until 1974.

In Nazi Germany, some forms of reconstructive surgery were made mandatory. In 1936, a law was passed that allowed the state to transform (*umgestalten*) a soldier's body without his consent, if this would benefit competence and power in combat. In Fascist Italy, Mussolini used surgery to improve his officers' physical charisma.

In these examples, the dominant group fanatically imposes their will upon those who deviate from the norm. This mechanism is extremely powerful because, as mentioned above, *homo sapiens* is a social animal, and therefore strongly inclined to accept the group norm. Even though humans can propose and cultivate a new standard as a form of resistance, such acts of defiance will usually result in a new group that accepts the new norm. There are very few persons who want to be, or can be, 'different' purely as an individual. To name just one example, radical changes in the fashion scene are still often accompanied by public outrage. We do not like being 'different'.

Fiddling with Age

Everybody wants to grow old and live a long life, but nobody wants to be old. This was true a century ago, and it still is. Many have searched for a way to stay young, but so far the fountain of eternal youth has eluded us—although through the years several researchers have claimed to be very close to its discovery. The Paris-based Russian surgeon Serge Avramovitch Voronoff (1866–1951), for example, could almost taste eternal youth. He implanted monkey testicles in older men (he did not use human testicles, because there were no young men willing to give them up; the apes were not too pleased with the procedure either, but their feelings were not considered). Voronoff noted a clear rejuvenation in his 'patients', but whether his research subjects agreed remains unknown.

In the US, Eugen Steinach (1861–1944) conducted a similar experiment. He believed that the secret of eternal youth was contained in a specific type of testicular cell, which he dubbed the 'puberty gland'. By activating these cells, youth would once again flow through elderly bodies. Max Thorek (1880–1960) tried to reactivate the bodies of several men by implanting them with monkey glands. Unfortunately, there is no information available about the effects of these experiments, but they are unlikely to have been very spectacular. They were methods invented for anxious old men, by anxious old men.

A variant on Voronoff and Steinach's methods was used to 'cure' homosexual men. These men were castrated, after which their testicles were replaced by those of 'healthy' heterosexual men. It is hardly surprising that the desired result was not achieved.

Body Images

Many of the concepts I have described above are closely related to body images, and as we have seen these are by no means stable. Nowadays, our favourite image is that of an eternally young, strong and good-looking man or woman. A beginning paunch is seen as a personal failure, and as a consequence 40-somethings can be seen running through the suburbs of Western cities every evening, wearing colourful tracksuits and iPods.

In his autobiography *The World of Yesterday*, Stefan Zweig describes how the streets of early-twentieth-century Vienna were still peopled with big-bellied men. Distinguished faces, heavy suits, hats, canes, cigars and, last but not least, rotund bellies. These figures strolled slowly, talked measuredly and stroked their carefully modelled, often greying beards. Grey hair was a new sign of dignity. A respectable man avoided haste and would never run. Haste and physical exercise were regarded as vulgar. The paunch and the heavy, bulky body had status and reflected the contentment of the bourgeoisie (see also Mulder 2005).

While nowadays tanned and well-toned bodies have become the ideal, strived after in sun studios or those body modelling factories called 'fitness clubs', at the end of the nineteenth and the beginning of the twentieth centuries such bodies were seen as evidence of a lack of prosperity. Only farmers and labourers were tanned and muscular, because of the time they spent outdoors performing physical labour. The poetic admiration of a woman's milk-white skin and snow-white thighs is well-known. Of course, there were always those who held a different opinion. Around 1900, several apostles of the body culture claimed that there was a direct connection between physical health, a muscular body and mental and sexual well-being (see Mulder 2005; Dekkers 2006).

The ideal shape of the human body is strongly dependent on the contemporary body image. The buxom nudes of Ingres would be considered fat in the current Western world, and the same applies to the three graces of Rubens and the bathers of Courbet. They no longer meet the current body ideal in any way.

At the end of the nineteenth century, an American newspaper printed an advertisement (Fig. 3.1) that encouraged women to buy Professor Williams' products, which would help them to 'end despair' and gain several pounds in just a few weeks. This advertisement had the catchy title of 'Respectfully Tell the Ladies Use "Fat-ten-U" Food to Get Plump'. In today's society, with its almost compelling demand of slenderness, such an advertisement is unthinkable—yet, there are only a few generations separating then from now. It is important to keep in mind, however, that there are still major cultural differences with regard to the ideal body type, especially for women. In the Arab world, a different ideal applies than in West European and Anglo-Saxon culture. The same goes for parts of Asia.

Nowadays, we want to be forever young, so advertising brochures tell us. Forever young to avoid the horror of Alzheimer's; forever young to be able to ignore the Zimmer frame. We need to take pills to prevent the decline of memory, we need to eat healthy food and we need to exercise.

Thus, Gilman's concept of 'passing' has gradually gained another meaning. In addition to the desire not to draw attention, to be average and to blend in with a group, the term now also comprises the individual desire to stand out, to remain young and to belong to a certain, otherwise unattainable beauty category. It is plausible that this development is connected with the societal pressure to conform to a certain standard of appearance. It is little surprising that this pressure exists— it being the socio-psychological essence of a group—and it is also well-known that individuals find it difficult to ignore this pressure. In the United States, the number of cosmetic surgical procedures in the category of rejuvenating and beauty operations increased between 1997 and 2000 by 173 %.

However, the use of cosmetic surgical products is not—yet—distributed normally across the population. Besides the financial hurdle, which can be quite substantial, there also appears to be a connection between plastic surgery and personality types. An empirical study on this subject conducted by Davis and Vernon shows a significant correlation between fear of abandonment and the desire to undergo a cosmetic surgical procedure. The greater this fear, the more willing a person is to undergo an operation to be, become or stay more attractive. However, the study by Davis and Vernon was conducted mainly among subjects younger than 30. Much remains to be said on the subject, as well as on the psychological effects of cosmetic surgery, but a full discussion is beyond the scope of the present chapter.

We have become the engineers of our new bodies, shaping them according to our own views. We cherish our bodies, so as to avoid death and diseases. I once overheard a man who had never smoked and had always exercised say that he felt

3 Changing the Body Through the Centuries

Fig. 3.1 Advertisement from the end of the nineteenth century

betrayed by his body when he was diagnosed with cancer. And that is exactly what he meant: betrayal, a broken promise. We are badly equipped to cope with the random play of chance, those few moments of oxygen deprivation during sleep that can alter a life entirely, that hidden tumour growing in the body, that rare type of cancer.

We are able to bend everything to our will, so why not control our bodies as well? This is why we take it bungee jumping, to show it who is in control: 'mind over matter'. This is why we take it to the gym, to shape it according to our ideal body image. And these are still relatively innocent aspects of the struggle to control the body; others are a great deal more radical.

The Body as a Work of Art

The French performance artist Orlan uses her body as a canvas to paint on. Since 1990, she has been working on the transformation of her own body. She does not use brushes, however, but knives, wielded by a surgeon who is the cutting extension of Orlan's will. In a series of nine operations, a plastic surgeon modelled her face on five mythical women from art history: Mona Lisa, Diana, Psyche, Europa and Venus. These examples of body art were broadcast live, and were acted out as if they were public performances. The audience was able to see, for example, how the scalpel cut through Orlan's lips. Her employees and the surgeons were dressed in designer outfits, and the operations took place in a fanciful decor.

Furthermore, the operating room was decorated with life-size images of Orlan and her muses. There were male striptease dancers, and during the procedures Orlan recited philosophical, literary or psychoanalytical texts. She also sold parts of her body. For a considerable amount of money one could buy bits of tissue encased in plastic, fat cells taken from her face and little pieces of scalp (Halsema 2007). Orlan actively interferes with her own body: she practises 'carnal art'. She literally takes the right given us by Giovanni Pico della Mirandola in the fifteenth century into her own hands by sculpting her body as if it were a work of art.

Orlan is convinced that our bodies are inadequately equipped for present-day technological possibilities. According to her, the body has to be reinvented. She believes that human beings will become increasingly capable of using and managing their bodies as vehicles, and considers biotechnology, psychotropic drugs and cosmetic surgery as the perfect tools to accomplish this goal. She sees her work as 'a fight against nature and the idea of God' (qtd. in Gilman 2000, p. 323). However, in 'Intervention' she claims: 'My work is not against cosmetic surgery, but against the standards of beauty, against the dictates of a dominant ideology that impress themselves more and more on feminine … and masculine flesh' (Orlan 1998, p 324).

While Orlan regards the body as a work of art, others are more interested in interfering with the body's design. Although this may also involve an individual's ideas about beauty, this category does not easily fit in with the generally accepted standards regarding the human body.

More information, pictures and wishes of individuals who belong to this group can be found online under the search term 'extreme body modification'. They differ from most other users of cosmetic surgery in that they are not interested in

3 Changing the Body Through the Centuries

inconspicuousness or in blending in with the anonymous mass, but rather in the opposite: the deliberate mutilation of the body. Examples include the *nullo*, the man who had his genitalia removed, and the woman who could no longer bear to live with legs. In an interview, she expressed her wish to have them amputated, as they did not feel like part of her body. Eventually, a hospital in Scotland was found willing to amputate her healthy legs. In some other cases surgeons refused, and patients decided to amputate the unwanted body parts themselves—the consequences were horrible. These individuals suffer from a body that does not seem to be represented in the brain and is thus perceived as alien and hostile. In psychiatry, this condition is called 'body dysmorphic disorder'.

In reality, the situation is more complicated. The term 'psychiatry' suggests that there exists a clear line between 'normal' and the pathological. A woman who chooses to amputate her perfectly healthy legs can be easily regarded as a psychiatric patient, but what about the healthy young man who decided to have an incision in his skull across the length of his forehead and to have his eyebrows removed? He wanted to change his appearance; he created a new person out of his old self—on the outside, that is. Is he a patient as well, or is he a body artist who recreates himself according to his own views? How does his case differ from Orlan's? The answer to this question is not straightforward.

Orlan reconstructs her body by means of knives, but there are also chemical ways to achieve this goal. In an interview with the Dutch newspaper *NRC Handelsblad* (issue of April 1, 2005), architect and Pritzker Architecture Prize winner Thom Mayne points out the enormous muscles of a woman in a picture. He indicates that this would have been impossible 30 years ago, since the chemicals she had to take in order to achieve this result were not yet available or would have killed her. Nowadays, building your body has become a part of general culture. Everyone can decide for themselves what their bodies should look like, believing they can create their own reality. Mayne derives a certain sense of optimism from the unlimited possibilities of human enhancement.

In some respects, the reverence for the body has disappeared. The body is no longer seen as the temple of the Holy Ghost, and no one cares about the threats made by Innocent III about immortal worms gnawing at our flesh. The body has become a machine in which we live and which has to take us from one place to another. The media display the body in all its shapes, both dressed and undressed. Medical programmes show the decline of the body and introduce viewers to aspects of physical reality which they would normally have never known. Violent deaths in crime shows are broadcast in colour, and makeover programmes teach us that our body is our property, with which we can do whatever we want. And in a sense they are right; the body is our property: if *I* am not the owner of my own body, then who is? There is much to discuss regarding the question of what we are allowed to do with this property, legally, ethically and theologically, but I will leave that for others to debate.

Human Enhancement

On this journey through time, several things have become clear. First of all, while the term 'human enhancement' has become more realistic due to the technological developments of the past decades, attempts to change the body have existed for a much longer time. For many centuries, humans have tried to intervene in the random fate dealt by nature's hand (deformities and diseases) or in the consequences of human actions (accidents and wars). For centuries, they have tried to reconstruct missing body parts, to shrink those that are too big and to enlarge those that are too small. Second, it is surprising to discover that in most cases these changes are not made in order to achieve a certain beauty ideal, but in order to blend in. Human beings are born adapters. In recent years, this concept of 'passing' has received a different connotation because of society's focus on youth and health, resulting in standards of appearance that differ significantly from those in the older examples, such as those given by Gilman.

At the moment, a new category is emerging, the hyper-individualists, who have the courage to step outside the established order. Current technology has provided them with new possibilities and has empowered them to genuinely and radically interfere with their body's design. This leads to fascinating results. Human beings want to be their own rulers and have perfected this control with the aid of technology. Therefore, it will be a long time before the last word has been said and written on human enhancement.

Bibliography

Camporesi P (1988) The incorruptible flesh: bodily mutation and mortification in religion and folklore. Cambridge University Press, Cambridge

Corbin A (1988) The foul and the fragrant: odor and the French social imagination. Harvard University Press, Cambridge

Davis K (2003) Dubious equalities and embodied differences: cultural studies on cosmetic surgery. Rowman & Littlefield, New York

Davis D, Vernon ML (2002) Sculpting the body beautiful: attachment style, neuroticism and use of cosmetic surgery. Sex Roles 47(3–4):129–138

Decety J, Batson CD (2007) Interpersonal sensitivity: entering others' worlds. Psychology Press, New York

Dekkers M (2006) Lichamelijke opvoeding. Contact, Amsterdam

Eco U (2004) History of beauty. Rizzoli International, New York

Featherstone M (1999) Body modification. Sage, London

Gilman SL (2000) Making the body beautiful: a cultural history of aesthetic surgery. Princeton University Press, Princeton

Hakman ECJ (1993) Een nieuw gezicht? diss. Amsterdam (VU). Bohn Stafleu Van Loghum, Houten

Halsema A (2007) Grenzen aan de maakbaarheid. Wijsgerig Perspectief 47(2):17–27

Huizinga J (1919/1996) The autumn of the middle ages. University of Chicago Press, Chicago

Labrie A (2001) Zuiverheid en decadentie. Over de grenzen van de burgerlijke cultuur in West-Europa 1870–1914. Bert Bakker, Amsterdam

3 Changing the Body Through the Centuries

Mann H (1918/2004) The loyal subject. Continuum International Publishing, New York

Meltzoff A, Decety J (2003) What imitation tells us about social cognition: a rapprochement between developmental psychology and cognitive neuroscience. Philosophical transactions of the Royal Society of London 358, series B, pp 491–500

Meltzoff A, Moore M (1983) Newborn infants imitate adult facial gestures. Child Dev 54:702–709

Mulder T (2005) De geboren aanpasser. Over beweging, bewustzijn en gedrag, Contact, Amsterdam

Mulder T (2007) Motor imagery and action observation: cognitive tools for rehabilitation. J Neural Transm 114:1265–1278

Nooteboom C (2001) All Souls' Day. Picador, London

Orlan (1998) Intervention. In: Phelan P, Lane J (eds) The ends of performance. New York University Press, New York, pp 315–327

Pico della Mirandola G (1965) On the dignity of man. The Bobbs-Merrill Company, Indianapolis

Schnabel P (ed) (2005) Hier en daar opklaringen. Sociaal en Cultureel Planbureau, The Hague

Thornhill R, Gangestad SW (1999) Facial attractiveness. Trends Cogn Sci 12(3):452–460

Valentine T, Darling S, Donnelly M (2004) Why are average faces attractive? The effect of view and averageness on the attractiveness of female faces. Psychon Bull Rev 11(3):482–487

van Campen C (2005) Een gelukkige geest in een cosmetisch lichaam. In: Schnabel P (ed) Hier en daar opklaringen. Sociaal en Cultureel Planbureau, The Hague, pp 42–45

Zweig S (1964) The world of yesterday: an autobiography. University of Nebraska Press, Lincoln

Chapter 4
Human Enhancement in Futures Explorations

Lucas Cornips and Marjolein van Asselt

Abstract Human enhancement is a common theme in fiction and popular culture (i.e. science fiction films) and often constitutes an attempt at exploring the future. This chapter sets out to what extent it is also encountered in more 'serious' and scientific explorations of futures. There are different approaches to the exploration of futures: long-term explorations, essayistic reflections and diagnoses of our time. These investigations originate in the desire to be able to anticipate the impact of new societal developments, unknown mechanisms and unusual circumstances. The exploration of futures is now an officially recognised research area and has inspired a novel branch of scholarship, comprising a multitude of different approaches and opinions and exploring a wide variety of possible future scenarios. However, human enhancement is a topic that is usually treated only marginally in such studies. Ultimately, most explorers of futures do not believe that human beings will be 'created', but they do stress the importance of trying to visualise the possible future developments and scenarios which could emerge from new science and technologies.

> Everyone says Brave New World is supposed to be a totalitarian nightmare, a vicious indictment of society, but that's hypocritical bullshit. Brave New World is our idea of heaven: genetic manipulation, sexual liberation, the war against aging, the leisure society.

Translated by Lucie Martin.

L. Cornips (✉)
Health Council of the Netherlands, P.O. Box 16052, 2500 BB, The Hague, The Netherlands
e-mail: lm.cornips@gr.nl

M. van Asselt
Technology and Society Studies, Maastricht University, P.O. Box 616, 6200 MD, Maastricht, The Netherlands
e-mail: marjolein.vanasselt@maastrichtuniversity.nl

B.-J. Koops et al. (eds.), *Engineering the Human*, DOI: 10.1007/978-3-642-35096-2_4,
© Springer-Verlag Berlin Heidelberg 2013

> This is precisely the world that we have tried—and so far failed—to create. (Houellebecq 2001, p. 131)

This statement is made by Bruno, one of the main characters in *The Elementary Particles*. In the epilogue to this novel, Houellebecq outlines a future scenario in which humanity takes full control over the conditions of its own existence. Biological evolution is no longer an elusive process, as *homo sapiens*, with all its shortcomings, is designedly turned into a new, immortal species. Houellebecq's epilogue can be considered as a backcasting scenario: the narrator looks back from a distant future on the transformation of society and the emergence of the enhanced human being. The scenario chronologically recounts 'how matters have got to this state' as the narrator describes the radical growth of biological knowledge to the point where technology allows us to control the genetic makeup of human beings. Moreover, the epilogue also deals with the required changes in attitude towards the creation of a new human race: the initial aversion against 'biological thinking' decreases and is steadily replaced by the realisation that the creation of a new species may even be desirable. The year 2029 is a historic year in this scenario, for it is then that the first new human being is created.

Discussions of human enhancement inevitably invoke visions of the future. Literary representations of enhanced humans are often set in a distant future, and introductions to the subject of human enhancement cannot ignore famous futuristic novels such as *Brave New World* by Aldous Huxley (1932), *1984* by George Orwell (1949), *The Island of Doctor Moreau* by H.G. Wells (1896) and *Frankenstein* by Mary Shelley (1818). Apparently, human enhancement is best imagined in the context of the future. Literary science fiction novels such as *Brave New World* and *1984* have become classics, familiarising their readers with the notion of human enhancement, but enhanced humans also figure in more recent futuristic novels, including *Oryx and Crake* by Margaret Atwood (2003) and *The Elementary Particles* (1998) and *The Possibility of an Island* (2005) by Houellebecq. In *Man in Progress: the Body as a Building Kit*, a collection of essays edited by Dorrestein et al. (2002), Dutch authors like Désanne van Brederode and Renate Dorrestein provide literary interpretations of various aspects of human enhancement, such as the impact of replacement medicine.

Over the past decades, several utopian and dystopian representations of human enhancement have reached an even wider audience through film and television. And these are not just screen adaptations of the literary science fiction novels mentioned above; popular science fiction films and television shows such as *Star Wars* and *Star Trek* also feature enhanced humans. *Star Trek's* 'Borg', for example, half-human, half-robot cyborgs, are known across the globe. Thus, representations of enhanced humans have worked their way into popular culture and are inextricably linked with the future.

This observation begs the question of whether human enhancement can also be found in more 'serious' and 'scientific' explorations of futures. And if this proves to be the case, what can futures studies teach us about the notion of human enhancement and the enhancement of human beings? These questions will be addressed in this chapter. First, we will provide a brief introduction to the field of futures exploration, after which we will examine how human enhancement is represented in a number of Dutch futures studies. We will adhere to Lüthy and Koops's description provided in the Introduction to this volume. Lüthy and Koops argue that human enhancement is a collective term which encompasses a wide range of technological applications, including cloning, gene therapy, cosmetic surgery, neurosurgery, cybernetics and nanotechnology. The only aspect these different technologies have in common is that they intervene in what is considered 'original' human nature. Lüthy and Koops distinguish five kinds of enhancement, four of which we will be investigating: (1) enhancement of the existing functions; (2) methods of regulation and selection in the reproduction of human individuals; (3) replacement or extension of natural by artificial human body parts; and (4) methods for influencing and controlling behaviour.

In addition, our contribution will supplement Lüthy and Koops's description, and on the basis of our analysis we will argue that the phenomenon of human enhancement is broader than a mere combination of biomedical interventions in the human body. In the last part of this chapter we will reflect on our results, relating our conclusions to Houellebecq's *The Elementary Particles* as well as Paul Schnabel's essay '2010 in Sight'. Finally, we will consider the question of whether futures explorers should consider human enhancement as a real challenge for the future.

Exploring the Future

Many companies, ministries and institutions engage in futures research, especially in the Netherlands. The goal of futures explorers is to provide insight into possible futures of the world or parts thereof, preferably on the basis of scientific knowledge about the past and the present. Typified as 'experts of promises' (see van Lente 2000), they systematically attempt to invent and reflect on probable, possible, desirable and undesirable futures. In the Netherlands, policy-oriented futures research is institutionalised to a high degree. Many Dutch institutions and organisations have the role, task, assignment or ambition to explore futures, including independent government planning agencies such as CPB (Netherlands Bureau for Economic Policy Analysis) and RIVM (National Institute for Public Health and the Environment), institutes for sustainable energy transition and other specialised consultancies such as Futureconsult (see futureconsult.nl). Advisory bodies also explore futures and are, on occasion, commissioned by ministries or the government. In addition, a large variety of other parties are active in the field: local authorities such as municipalities, provinces and cities; further government bodies; and scientific departments of political parties, committees and foundations (e.g. the STT Netherlands Study Centre for Technology Trends). Lastly, futures are also explored by companies, including the Dutch energy company Essent and Dutch cooperative bank Rabobank; individual authors such as the Das brothers, two well-known Dutch futurists; and researchers at universities.

There are different types of futures explorations. In their analysis of Dutch futures studies, van Asselt et al. (2005) propose the following categories:

- *long-term studies* in which substantiated accounts are given on what the future might look like;
- *essayistic reflections* that contain or represent ideas about the future;
- *diagnoses of the present*, which explore structural developments of the past and the present and place relevant topics on the agenda for the future.

It is questionable to what extent diagnoses of the present should be considered true futures studies. For this reason, this chapter will only deal with long-term studies and essayistic reflections. The distinction between these latter two approaches is also blurry, however, because some long-term studies are based on essayistic reflections. A case in point is *Polar Bear Plague on the Veluwe* edited by in 't Veld and van der Veen (2006), a collection of essays which will be discussed later. In long-term studies, two types of statements regarding the future can be distinguished: (1) prognoses and (2) scenarios. A prognosis or 'prediction' is a point estimate that is generally based on the extrapolation of past trends, sometimes including a range of possible deviations. The underlying assumptions are that no other processes will interfere, that important mechanisms are sufficiently understood, and that no discontinuities will occur, such as trend breaks, surprises or structural disruptions. Although many futures studies do anticipate change in the future, these changes are often gradual; it is a question of 'more' or 'less' of a particular trend or development (see van Notten 2005).

In many cases, however, explorers study futures in order to consider different circumstances, new developments, unknown mechanisms and unusual combinations of events, and in order to do so they create futures scenarios. The basic principle of working with scenarios is that futures cannot be predicted and that it is therefore essential to take into account various possible futures. Generally, these are the result of thought experiments: 'what would happen if…?' The answer to this question can be given as a 'final image'—that is, a description of the future at a specific moment in time (e.g. the year 2030) or as a 'scenario plot': a description of a sequence of events and developments over a certain period of time. Explorers of futures who employ the latter method can thus be compared to script writers, and while a scenario plot is similar to a film, a final image resembles a photograph. In our analysis of Dutch futures studies from the perspective of human enhancement, we have included both futures essays and futures scenarios, as well as final images and scenario plots. In all cases, the selected futures studies are taken seriously in social and scientific debate.

Technological Revolutions and Designer Babies

Does human enhancement feature in futures explorations? And if so, how is it portrayed? It is useful to start with van Steenbergen's futures essays 'Man on the Throne of God?' (2002) and 'The New Human Being in a Future World Society'

4 Human Enhancement in Futures Explorations

(2003), as he explicitly depicts the 'new' human being as a major challenge for the future. Human enhancement, van Steenbergen expects, will be made possible in the coming decades through a biomedical revolution that causes medical science to move from a curative (past) and preventive style (present) towards a design-based approach. He describes how the resistance to viewing 'normal' human traits as needing treatment and improvement will decline in the future under the influence of a current trend: the rise of 'biological thinking', also known as 'biologism'. According to van Steenbergen, the familiar statement 'it's in my genes' is a perfect example of this trend. He expects that biologism will continue to grow and will thus pave the way for the acceptance of genetic interventions, and argues that human enhancement is made possible by a combination of technologies:

> On the basis of developments in the field of cloning, DNA-research and eugenics, combined with those in the field of information technology, a new stage known as the 'design phase' is to be expected in the not too far future. In this context, the term 'designer child' is used. (van Steenbergen 2002, p. 695)

Moreover, Van Steenbergen (2003) discusses the notion of 'designer babies' and a 'design phase' in medicine. These mechanical metaphors have an important rhetorical effect: they suggest the possibility of human enhancement. Van Steenbergen mainly focuses on methods of selecting desired characteristics, but the improvement of the existing functions is also discussed in his essay.

Hendriks' futures essay 'The Engineerable and Self-mutating Human' (2006) is another exploration in which human enhancement is explicitly discussed. This essay is a contribution to the futures exploration *Polar Bear Plague on the Veluwe*, a collection of essays about the future, including the future of human enhancement. The title of this collection refers to an unlikely situation, the Veluwe being a natural forest area in the Netherlands, and was selected to acknowledge the possibility of discontinuities in the future: 'it is a tribute to the unpredictability of the future' (in 't Veld and van der Veen 2006, p. 11). Hendriks describes human enhancement as follows:

> Some believe that scientific developments, especially in biotechnology, will enable human beings to steer human development and evolution in a desired direction; that even more health, beauty, intelligence, vitality, power, and 'eternal' life will become attainable. (Hendriks 2006, p. 206)

Hendriks discusses several technological developments, such as nanotechnology, gene therapy and regenerative medicine. Unlike van Steenbergen, he does not mention a revolutionary integration of different human enhancement technologies. Hendriks' essay provides an extensive list of future technologies, encompassing three of the five types of enhancement: (1) enhancement of functions, (2) selection of characteristics and (3) replacement of human body parts. In addition, he describes a number of possible applications for various technologies, including nanobots (that is, robots on the nano scale) that can travel via the bloodstream to a diseased cell in order to cure it; genetic medicines developed through gene therapy in order to cure rare hereditary diseases; and organs such as kidneys, which could be created using regenerative medicine. The latter might enable life prolongation because any damage caused by ageing could be reversed.

Furthermore, Hendriks draws attention to one particular problem of human enhancement: inequality. He believes that opportunities for enhancement will not be accessible to all. Since healthcare will be modified and will take a more individual, rather than collective, approach, healthcare costs will increase. Thus, solidarity will be put to the test, and health insurers will differentiate their premiums. As a result, the less fortunate will hardly have any access to the technologically high-quality, but very expensive, healthcare system.

Smart Pills

Hendriks' futures essay is unique among the futures explorations investigated for this essay in that it explicitly mentions human enhancement. Van Steenbergen (2003), for example, merely alludes to 'the new human'. This does not mean, however, that the concept of human enhancement is absent in other futures studies. Van Santen, Khoe and Vermeer's *Smart Pills* (2006), a futures exploration conducted at the Dutch Eindhoven University of Technology, contains several scientists' perspectives on the future of their respective fields. The design of this study prevents an exploration of human enhancement as a product of converging technologies, for each chapter covers a different technological development.

In their treatment of these various technologies, the authors avail themselves of two interesting strategies. First of all, the entire collection is characterised by the use of technological terms for human body parts and body characteristics. Parts of the body are invariably designated with mechanical metaphors: proteins are factories, brains are computers and human DNA is an operating system like Windows. In the chapter entitled 'Eternal Life', it is explained that human beings do not have a 'built-in time bomb' that will inevitably kill them once they have reached a certain age. Instead, nature pays little attention to long-term maintenance, and we die as a consequence of 'overdue maintenance problems', rendering us a 'total loss' (p. 189; see also Chap. 8, this volume). These metaphors are relevant to the topic of human enhancement, because they suggest that humans can be 'built' just like machines. The use of mechanical metaphors is a tried-and-tested method or strategy to suggest a certain amount of control over and 'engineerability' of the described object (see Hayles 2004). Moreover, this choice of terminology also raises questions about humanity: if humans can be compared to machines, what does that imply for our understanding of the human condition?

Besides mechanical metaphors, a second notable strategy of this collection involves the depiction of technology as derived from nature. Technologies cannot be seen as 'artificial' because they are based on 'intelligent' nature. It is argued that 'nature started to think of smart solutions one billion years before humans did, and we can learn from that. It pays to try to imitate nature and, if necessary, to improve upon its natural processes' (p. 154). This emphasis on imitating 'intelligent nature' is interesting in the light of Lüthy's conceptualisation of human enhancement. In his historical–philosophical contribution (Chap. 2, this volume),

Lüthy argues that enhancement is associated with the manipulation of what is considered the natural human condition. In the past, something that had been 'enhanced' was perceived as *superior* to nature. As far back as two millennia ago Plato already expressed his wish to combat all 'limitations' of the natural human being by means of science—that is, artificial intervention. Human domination over nature, Lüthy describes, was an important factor in the realisation of a 'well-bred' human being. It is notable, therefore, that in *Smart Pills* nature is represented as superior to humans and is depicted as a model for enhancement technologies. In this study, the aim is not to rise above the inferior world of nature by means of technology, but rather to learn from 'intelligent' nature and to use the acquired knowledge to improve technology and, by extension, the human race.

Another interesting aspect of *Smart Pills* is its emphasis on the necessity to check for deficiencies. The chapter entitled 'Smart Pills', for example, indicates that in the future—despite progress in biomolecular medicine—defects will have to be detected at an early stage. Everyone will need to monitor themselves often, and some may even be required to purchase a scanning device as electronic protection. In the chapter 'Electronic Protection', progress in the field of diagnostic methods for identifying heart failure leads to devices that can be used to detect cardiac arrhythmias at home. Lastly, the chapter 'Renewing the Body' argues that while techniques for creating artificial organs will significantly improve in the future, replacement technologies will be especially successful where the healing of small parts of organs is concerned. Therefore, it will become even more important to detect abnormalities at an early stage and to undergo frequent scans or other screening tests. In addition to Lüthy and Koops's five types of enhancement, *Smart Pills* mentions a sixth: technologies to detect 'anomalies' and to control and analyse a variety of bodily functions. At the same time, detecting, controlling and monitoring are important conditions for enhancement, for without knowledge of their deficiencies, human beings cannot be enhanced.

In *Smart Pills*, enhancement is mainly associated with the replacement of human body parts by technological artefacts, such as a new heart valve, the replacement of parts of organs or the implementation of a hearing aid. Improvement of the existing functions only recurs *in abstracto* in the aim to improve and prolong the life of the imperfect human through technological means. The study mentions only a few concrete examples of enhanced humans, because technologies are usually discussed in the abstract and not in the context of everyday life. In other words, it is technology itself—rather than its effects and interaction with humans—that is at the heart of *Smart Pills*.

Experience 2030

Human enhancement is represented in an entirely different way in Essent's *Experience 2030* (2003). This book offers a very broad perspective: major social developments as well as the future of the world are main themes of this futures study. Possible social developments are worked out in four scenarios set in the year 2030. The scenario 'Hyper-individualism' describes a future in which technology has profoundly changed human life at both the individual and the societal level. Reprogenetics—the use of genetic technologies in reproduction—has given parents the opportunity to exercise full control over the lives of their children both before and after birth, and technology is even deployed to alleviate sadness and loneliness. Due to these technological developments, society has become highly individualised. Moreover, this is not the only adverse side effect of human enhancement: some individuals also feel a longing for the past, when life was not yet so 'pre-programmed'. They experience a lack of dynamics, tension, uncertainty and excitement. In this scenario, the enhanced human is a product of all four types of enhancement discussed in this chapter and is best characterised in the comparison to God: '[there is] no distinction between God and engineers anymore. Engineers are God. They can and will do anything' (p. 75). Man is no longer a product of divine creation, but a product of engineering.

Another scenario in *Experience 2030* introduces a dichotomy between what we might call enhanced humans, the *haves,* and the poor, non-enhanced *have-nots.* The *haves* are technologically very advanced, and their lives are shaped by technology. There is a strong belief in technological solutions for problems at both the individual and the collective level, and anyone who can afford the technology can eradicate practically all their physical and mental imperfections. Individuals have chip implants to monitor their health. Like *Smart Pills*, this scenario shows how knowledge of the individual is a necessary precondition for human enhancement. *Preview 2030* (Essent 2004), a futures exploration containing further reflections on the scenarios sketched in *Experience 2030*, gives a thought-provoking example of the future belief in the genetic basis of human beings. *Haves* who wish to work for a company have to include a genetic passport in their application, and on the basis of this passport the company will judge whether or not they are suitable candidates. In other words, the genetic passport certifies whether the applicant's talents and health meet the company's requirements.

Disciplining Behaviour

In van Steenbergen's futures essays and in the future scenario 'Hyper-individualism', the enhanced human is perceived as the result of an integration of several types of enhancement. In other essays and scenarios dealing with human enhancement, enhancement technologies are treated in isolation. So far, we have discussed futures explorations that primarily focus on physical interventions, but Lüthy and Koops also include methods of influencing and controlling behaviour in their list of enhancement technologies. It should be noted, however, that human behaviour can be controlled and disciplined in other ways besides technological interventions in the human body. If we were to include disciplined behaviour in our conception of the enhanced human, we would encounter examples of human enhancement in many other futures studies.

An established way of disciplining humans by means of technology is to monitor them with cameras. This enhancement technique is described in *Inside the Domestic Sphere* (Koops et al. 2004). In recent years, camera surveillance has been increasingly used in public areas, and if this trend perseveres—in combination with improved observation technologies—it will become possible to control, and therefore discipline, human beings even more efficiently in the future. The authors of this futures exploration expect that by linking cameras we will be able to monitor each person individually. It is even conceivable that researchers will develop cameras that are able to detect and register individuals' chemical characteristics. This form of technological behaviour enhancement is also mentioned in Essent's 'Haves and Have-nots' scenario (2003). Here, control is directed at the *haves* who live in secure compounds:

> The corridor outside the compound is continuously scanned for suspicious movements with infrared cameras. All incoming traffic is subtly monitored. Although less overtly perceptible, the control is more meticulous and more effective than that in the former Eastern Bloc, thanks to technology. (Essent 2003, p. 133)

Monitoring by means of technology forces the *haves* to behave according to the rules; they are being disciplined. However, this monitoring and disciplining of behaviour cannot be exclusively interpreted as a technique to control the *haves*, as it also affects the *have-nots*. Due to technological monitoring, *have-nots* are effectively shut out and excluded from all benefits of the compounds, including the technological ones. This future scenario shows how inequality can be created and maintained through enhancement technologies. There is a gap between the 'lucky' recipients of various technological and biological interventions and the 'unfortunate' individuals who are denied access to these forms of enhancement. In this scenario, the gap between the two groups expands with each generation because the *haves* are able to genetically enrich themselves, a form of enhancement that is not at the disposal of the *have-nots*. In other words, 'Haves and Have-nots' raises questions about the unequal distribution of, and resulting from, enhancement.

Enticing Behaviour

Behavioural monitoring and controlling methods are not only used to discipline the individuals observed. In Philip Idenburg's scenarios 'Marketing in Times of Growth' (2005) and 'Marketing in Times of Survival' (2005), technology is deployed to gather information about consumers' purchasing behaviour. In the future scenario 'Marketing in Times of Growth', individuals are observed in shops. Walking routes, shopping times, purchases, interests and even eye movements are recorded, and this personal information is used by commerce to develop individualised and highly aggressive sales methods based on customised customer enticement. Furthermore, the scenario also describes a future of technologically advanced and highly individualised medicine. Philips has strongly developed the 'life sensor' market. Individuals can buy chips and sensors that can be implanted into their own bodies, and by means of a personal digital assistant (PDA) their health can be continuously monitored.

In Essent's 'Hyper-individualism' scenario, energy company High Energy strives to obtain comprehensive knowledge of individuals, asking various research institutes, service providers and sales organisations to report on their habits. This allows High Energy to crawl into their clients' skin and to entice them individually, just like the customers in Idenburg's 'Marketing in Times of Growth'. Each customer is presented with a 'hyper-personal' package of services, in which various options are adapted to their personal demands. This package is offered under a separate brand name, so that each client feels they are offered an individual and exclusive treatment.

In 'Marketing in Times of Survival', behavioural information is not only used to monitor individuals' purchasing habits, but data from the vast databases are also used in improper ways:

> Abuse of customer information for companies' personal interests (the unauthorised storage and linking of files) is the order of the day. Blackmail and 'preferential treatment' based on database data (e.g. medical, criminal record, or financial) are increasingly common. (Idenburg 2005, p. 102)

In these futures scenarios, monitoring is deployed to entice customers—a more indirect yet very effective way of controlling behaviour. The marketing scenarios are especially interesting, because they also question whether it is desirable to accumulate extensive knowledge of individuals and their habits. The futures explorations mentioned so far amply discuss the monitoring of behaviour and the accumulation of information about human characteristics. This knowledge is used to 'shape' individuals in various ways. In this way, enhancement technologies are combined with thorough monitoring of all sorts of physical properties. The marketing scenarios demonstrate that the acquired personal information can also be used in unacceptable ways.

The combination of information acquisition and personalised interventions features strongly in *Genomics 2030* (de Graef 2005). In this collection of futures essays, a number of authors outline various visions of the future of the world of genomics and society around the year 2030. In his introduction, de Graef argues that developments in medicine and sciences such as nanotechnology will cause healthcare to become more directed towards the individual. For example, pharmaceutical companies will be better able to adapt medicines to the needs of individuals, which will make healthcare more personal and effective. An adverse consequence of these developments is that healthcare will become more expensive. The idea of 'personalised medicine' is discussed in greater detail in Theo Verrips' contribution to *Genomics 2030*. Verrips claims that a personal approach will become possible through knowledge of an individual's genetic profile. Since it will be possible to 'compute' a person's risk of developing a certain disease, prevention will be central to this new system of medicine. Moreover, prevention profiles will render it possible to provide personalised advice on the best lifestyle, diet and use of medicines. This ideal of personalised and preventative medicine can be brought into practice through the combination of a knowledge system and a so-called 'virtual health agent':

> It will be a knowledge system that will be able to adapt the general knowledge system to a person's particular genetic make-up and lifestyle and communicating with him or her face-to-face using speech recognition. [...] Adaptation of one's lifestyle according to the recommendations of this agent can then be guided and monitored in such a way that the 'client' will never get the impression that 'big brother is watching me'. (Verrips 2005, p. 66)

A similar scenario is outlined in Groenewegen, Hansen and ter Bekke's *The Future of the Medical Doctor* (2007), a futures exploration concerning healthcare professions. In this vision of the future, humans are not only able to monitor their own health, but patients are also able to independently apply medical technology, as technology will become less complicated and therefore easier to apply. In addition to user-friendliness and autonomy of individual patients, this development will also lighten the workload of healthcare workers.

In Table 4.1, we have systematically summarised the various types of enhancement as well as certain particularities of the futures essays and futures scenarios mentioned in this chapter. All types of enhancement refer back to Lüthy and Koops's classification in Chap. 1.

The Limits of Human Enhancement

So far, we have discussed a number of futures explorations that reveal a strong faith in the possibilities of human enhancement, but in our research we also encountered futures studies that explore its limits. Some futures explorers attach importance to reflection on the *impossibility* of human enhancement. Gert-Jan van Ommen's contribution to *Genomics 2030* addresses this concern. He argues that genes are not the sole determining factor in human life and that knowledge of the human gene map will therefore not necessarily lead to control over human imperfections:

> To conclude, we have still a long way to go. Indeed, the closer we thought we would get to the 'original plan', predicting our future from our past, the clearer it becomes that we have all too easily overlooked the impact of the present: the interaction between our genetic makeup—unique for each individual——and the environment. (Van Ommen 2005, p. 30)

Naturally, this position has implications for the possibilities of human enhancement. According to this perspective, truly predictive medicine as described by van Steenbergen is not a realistic option for the future. Human beings are not solely determined by their genes, and therefore we cannot predict the future of their physical condition—let alone design it by such means as gene therapy. This is where human enhancement reaches its limits.

Table 4.1 Overview of human enhancement in futures explorations

Futures exploration	Dominant types of enhancement	Particularities	Possibly arising questions
Van Steenbergen: 'Man on the Throne of God?' (2002) 'The New Human Being in a Future World Society' (2003)	Integration of technologies, with emphasis on: – enhancement of existing functions (1) – methods of regulation and selection in the reproduction of human individuals (2)	Medical sciences will in the future have a design-based approach	
Hendriks: 'Engineerable and Self-mutating Human' (2006)	Enhancement of existing functions (1)	– Much attention paid to regenerative medicine – Enhancement as a product of diverse, separate technologies	Will healthcare become unaffordable for some?
Van Santen, Khoe and Vermeer: *Smart Pills* (2006)	Replacement of organs or human body parts by artificial elements (3)	– Use of mechanical metaphors – Nature as a model for enhancement technologies – Gathering knowledge as an important condition for enhancement	Do we agree with the image of human beings which informs the mechanical metaphors?
Essent: 'Hyper-individualism' (2003)	Combination of four types of enhancement	Society strongly individualistic and predictable	Does technological enhancement remove positive dynamics and excitement?
Essent: 'Haves and Have-Nots' (2003; 2004)	– Methods of regulation and selection in the reproduction of human individuals (2) – Methods for influencing and controlling behaviour (4)	– Inequality in human enhancement – Observation of behaviour; controlling and disciplining	What influence do enhancement technologies have on social inequalities?
Koops et al.: *Inside the Domestic Sphere* (2004)	– Methods for influencing and controlling behaviour (4)	– Observation of behaviour; controlling and disciplining	

(continued)

4 Human Enhancement in Futures Explorations   57

Table 4.1 (continued)

Futures exploration	Dominant types of enhancement	Particularities	Possibly arising questions
Idenburg: 'Marketing in Times of Growth' (2005) 'Marketing in Times of Survival' (2005)	– Methods for influencing and controlling behaviour (4)	– Gathering extensive knowledge on purchasing behaviour; customised enticement	Will knowledge of behaviour be used in unacceptable ways?
De Graef: *Genomics 2030: Part of Everyday Life* (*General Introduction*) (2005)	– No dominant type of enhancement	– Medicine individualised	Will medicine become unaffordable?
Verrips: *Genomics 2030*: *Part of Everyday Life* (2005)	– Enhancement of the existing functions (1) – Methods for influencing and controlling behaviour (4)	– Personal virtual health agent	
Groenewegen et al.: *The Future of the Medical Doctor* (2007)	– No dominant type of enhancement	– Personalised medicine; patients will be able to independently apply medical technology	

Another form of non-enhancement can be encountered in futures explorations that reflect on ways to influence human behaviour. Some explorations are based on the assumption that humans act according to fixed behavioural patterns, which are relevant because interventions will not be successful in such cases. For example, CPB, MNP and RPB's futures study *Welfare and Environment* (2006) discusses correlations between income and car use. In the background section of this study, it is stated as axiomatic that higher incomes lead to increased car use. Such rule-governed correlations constitute an example of the impossibility of human enhancement, as the presentation of behavioural patterns as laws leaves no room for enhancement by way of government measures or other forms of control.

A similar type of non-enhancement is described in RIVM's *Quality and Future* (2004). In the chapter 'Energy Supply', it is explained why it is so difficult—or even impossible—for citizens to save electricity. One reason in particular deserves special attention, because it shows a lack of belief in the possibility to enhance behaviour:

> Consumers' energy consumption is largely governed by 'hard' environmental factors such as income, family composition, housing, residential and work location and the limits imposed by money and time. Such factors can be influenced, but once choices have been made, consumers' freedom of choice becomes restricted for a considerable period of time.

In the small room for choice that remains, 'soft' environmental factors play a major role in the decision-making process. Such factors include education, socio-cultural background and beliefs. In addition, it should not be forgotten that many daily 'action choices' are not actual decisions but automatic behaviours. (RIVM 2004, p. 128)

'Energy consumption governed', 'freedom of choice restricted', 'automatic behaviours'—the authors of this futures exploration clearly see limits to the possibilities of human enhancement through behavioural interventions.

Reflection

In the futures studies analysed in this chapter, the belief in societal enhancement that was so prominent in the 1970s—and which also features in futures studies dating from that decade—has been replaced by an emphasis on the enhancement of individuals. Many of the described enhancement technologies are essentially individualistic in character. In his exploration '2010 in Sight' (2001), Dutch sociologist Paul Schnabel offers an interesting perspective on 'individualised enhancement'. According to Schnabel, enhancement has been redirected; since the concept of malleable society proved an illusion due to citizens' 'deviant and calculating' attitude, the notion of enhancement has shifted to the lives, persons and bodies of individuals. Life is increasingly perceived as a personal choice, a project that needs shaping. Due to the fact that 'shaping' your own person is a difficult task, assistance in this process has become a flourishing trade. Consultants, trainers and psychotherapists all contribute to the hyper-individual interpretation of human enhancement.

This image of hyper-individual human enhancement comes back in various futures studies, including *Smart Pills* (van Santen et al.), *Inside the Domestic Sphere* (Koops et al.), *A Vision of the Future* (Idenburg), the scenario 'Hyperindividualism' in *Preview 2030* (Essent), *Genomics 2030* (de Graef and Verrips) and *The Future of the Medical Doctor* (Groenewegen et al. 2007).

This shift in the focus of enhancement raises the question whether human enhancement is a typical phenomenon of our present day and age. In this respect, it is interesting to revisit the futuristic novel with which we opened this chapter: *The Elementary Particles* by Houellebecq. In this novel, humanity undergoes a radical transformation as a biological revolution breaks out, resulting in the emergence of a new kind of human being. However, the issue of hyper-individual enhancement does not figure in this 'scenario'; rather, the newly created species is stripped of all forms of individuality. In the epilogue, Houellebecq explains '[t]hat mankind must disappear and give way to a new species which was asexual and immortal, a species which had outgrown individuality, separation and evolution' (p. 258). Thus, Houellebecq provides an original and provocative counterpoint to the ubiquity of 'individualised enhancement' as is present in many recent futures explorations. In other words: *The Elementary Particles* is relevant to our contribution because, like Schnabel's reflection, the novel offers a fresh perspective on the dominance of 'individualised enhancement'.

As a source of inspiration for reflection on human enhancement, *The Elementary Particles* brings us back to the issue of 'fiction and the future'. In our introduction, we argued that human enhancement is inseparable from the future and that it is often represented in science fiction. In the introduction to *Polar Bear Plague on the Veluwe*, Roel in 't Veld and Hans van der Veen argue that futuristic novels may be valuable for a qualitative exploration of the future, and point out that literary and cinematic representations of the future can serve as a mirror. It is not only the visions' technological content that is interesting in this respect, but also and primarily their socio-cultural aspects.

While science fiction offers many representations of human enhancement, more serious explorations of the future do not. The futures explorations discussed in this chapter were selected from a large collection of futures studies, and it was difficult to find futures explorations that discuss the subject of human enhancement: most explorations do not address the issue of enhancement technologies. This may imply that futures explorers do not seriously envisage these developments and that such visions of the future only belong to the realm of fiction. However, it could also imply that futures explorers have a blind spot; that they, for some reason, overlook the theme of human enhancement and indeed need science fiction to become aware of its existence.

We are under the impression that not enough attention is being paid to human enhancement in futures studies, and that those explorations that do consider the issue generally lack a cohesive approach. Enhancement technologies are considered in isolation from their context while their social consequences are rarely taken into account, since it is the technology itself and not its impact on humankind and society that is the focal point of these studies. Moreover, a hyper-individualistic conception of enhancement usually predominates, even though futures studies are ideally suited to explore different interpretations and the boundaries of human enhancement. Even if futures explorers believe that humans will not be enhanced in the future, it is relevant for them to indicate which developments or factors will cause human enhancement to belong to the imaginative realm of science fiction. In his study *Promising Technology*, van Lente has shown that expectations created through futures scenarios have been used effectively to generate momentum and funds for technological developments. If serious futures explorers could refute certain futures scenarios, this would be a major contribution to the debate on human enhancement and enhancement technologies. Therefore, it is essential that they turn their gaze towards human enhancement, and we hope that this chapter will invite them to do so.

In addition, our contribution also provides clues for social and scientific debate on human enhancement. Those contemporary futures studies that do pay attention to the topic of human enhancement supply interesting themes for debate. Examples include the desirability of enhancement technologies, the possibility of abuse and manipulation, the increase of social inequality engendered by enhancement technologies, and the limits of human enhancement, as well as the question to what extent human enhancement is the new 'enhancement illusion'.

Bibliography

CPB, MNP, and RPB (2006a) Welvaart en leefomgeving [Welfare and Environment]. Een scenariostudie voor Nederland in 2040. CPB, MNP, and RPB, The Hague/Bilthoven

CPB, MNP, and RPB (2006b) Welvaart en leefomgeving. Een scenariostudie voor Nederland in 2040—Achtergronddocument. CPB, MNP, and RPB, The Hague/Bilthoven

de Graef M (ed) (2005) Genomics 2030. Part of everyday life. STT Netherlands, The Hague

Dorrestein R et al (2002) Mens in uitvoering. Het lichaam als bouwpakket [Man in progress: the body as a building kit]. Maarten Muntinga, Amsterdam

Essent (2003) Beleef 2030 [Experience 2030]. 4 toekomstscenario's voor de energiewereld. Essent

Essent (2004) Beproef 2030 [Preview 2030]. 4 toekomstscenario's in perspectief. Essent

Groenewegen P, Hansen J, ter Bekke S (2007) De toekomst van de witte jas [The Future of the Medical Doctor]. Professies en de toekomst: veranderende verhouding in de gezondheidszorg. VVAA, Utrecht

Hayles K (ed) (2004) Nanoculture. Implications of the new technoscience. The Cromwell Press, Wiltshire

Hendriks R (2006) Maakbare en zelfmuterende mens [The engineerable and self-mutating human]. In: In 't Veld and van der Veen (eds)

Houellebecq M (1998/2001) The elementary particles. Vintage International, New York

Idenburg P (2005) Oog voor de toekomst [A vision of the future]. Over marketing en consumenten in een veranderende samenleving. Scriptum Management

in 't Veld R, van der Veen H (eds) (2006) Polar bear plague on the Veluwe. Essays on the future. COS, The Hague

Koops BJ, van Schooten H, Prinsen M (2004) Recht naar binnen kijken [Inside the Domestic Sphere]. Een toekomstverkenning van huisrecht, lichamelijke integriteit en nieuwe opsporingstechnieken. Sdu, The Hague

Milburn C (2004) Nanotechnology in the age of posthuman engineering: science fiction as science. In: Hayles K (ed) Nanoculture. Implications of the new technoscience. The Cromwell Press, Wiltshire

Ministerie van Binnenlandse Zaken en Koninkrijksrelaties (BZK) [Ministry of the Interior and Kingdom Relations] (2005), Houdbaarheid verstreken. Toekomstverkenning en beleid, The Hague

RIVM (2004) Kwaliteit en toekomst [Quality and future]. Verkenning van duurzaamheid. Sdu, Bilthoven

Schnabel P (2001) De toekomst van het sociale domein. In: Rademaker P (ed) Met het oog op 2010 [2010 in sight]. De toekomst van het sociale domein, verbeeld in elf essays. De Balie Amsterdam, pp 111–120

van Asselt M (2000) Perspectives on uncertainty and risk. The PRIMA approach to decision support. Kluwer Academics, Maastricht University, Maastricht

van Asselt M, van der Pas J, de Wilde R (2005) De toekomst begint vandaag. Inventarisatie toekomstverkenningen. Onderzoeksrapport, Maastricht University

van Lente H (1993) Promising technology: the dynamics of expectations in technological developments. Twente University, Enschede

van Lente H (2000) Forceful futures: from promise to requirement. In: Brown N, Rappert B, Webster A (eds) Contested futures: a sociology of prospective techno-science. Ashgate, Aldershot

van Notten P (2005) Writing on the wall. Scenario development in times of discontinuity. Dissertation, Maastricht University, Maastricht

van Ommen G (2005) A short history of genomics. In: de Graef (ed)

van Santen R, Khoe D, Vermeer B (2006) Zelfdenkende pillen [Smart pills]. En andere technologie die ons leven zal veranderen. Nieuw Amsterdam, Amsterdam van Steenbergen B (2002) Man on the throne of God? The societal implications of the biomedical revolution. Futures 34(8):693–700

van Steenbergen B (2003) De nieuwe mens in de toekomstige wereldmaatschappij [The new human being in a future world society]. Uitdagingen voor de toekomstonderzoeker. Nyenrode University, Breukelen

Verrips T (2005) Genomics 2030: part of everyday life. In: de Graef M (ed)

Chapter 5
Genetic Enhancement of Human Beings: Reality or Fiction?

Annemiek Nelis and Danielle Posthuma

Abstract This chapter questions to what extent knowledge about the human genome may be used in order to plan or determine the future of an individual. While some human characteristics, such as the colour of the eyes or the hair, are indeed determined by genes, environmental factors also play a large role in the development of most of them, for example height. It remains largely unknown what precisely determines who we are or who we become. Where hereditary diseases are concerned, it is very rare that they are caused by a single gene and are thus relatively easy to detect. Usually, a multitude of genes interact in largely unpredictable ways with each other, as well as with the environment, to determine the outcome. Therefore, no definitive answers can be expected in the future from research conducted into multiple-gene disorders and hereditary characteristics, and assumptions about genetic determinism are, in most cases, misguided.

In the Hollywood film *Gattaca*, a society is depicted where individuals are selected for a career, marriage or social position on the basis of genetic aptitude. The film's protagonist, Vincent, aspires to a life as an astronaut. According to his genetic profile, however, he has a congenital heart condition, is short-sighted and has a life expectancy of 30 years—disqualifying him for a career in space. Nevertheless,

---

Translated by Aleid Wassens.

A. Nelis (✉)
Dutch Safety Board, P.O. Box 95404, 2509 CK, The Hague, The Netherlands
e-mail: a.nelis@safetyboard.nl
www.safetyboard.nl

D. Posthuma
CNCR, De Boelelaan 1085, room B-439, 1081 HV, Amsterdam, The Netherlands
e-mail: d.posthuma@vu.nl
www.ctglab.nl

B.-J. Koops et al. (eds.), *Engineering the Human*, DOI: 10.1007/978-3-642-35096-2_5,
© Springer-Verlag Berlin Heidelberg 2013

Vincent is determined to reach his goal and borrows the genetic identity of Jerome, a former swimming athlete who is genetically almost perfect but, after a tragic accident, is in a wheelchair and confined to his house. Despite being genetically imperfect, Vincent reaches his goal and becomes an astronaut because of his determination, the use of Jerome's identity and the expectations others have of him based on this identity.

Gattaca raises several interesting points, the first being genetic determinism: the belief that all human characteristics are entirely determined by genetic makeup. In Vincent's world, the recruitment of employees and even the selection of a possible partner are not based on conversations and meetings but on a comparison of DNA profiles. 'Show me your genes and I will tell you who you are' is the motto of this society. However, the film also rejects the idea of genetic determinism. Vincent may be using someone else's identity, but it is his own determination that enables him to become who he wants to be. This becomes clear in several scenes. Although his brother, parents and potential employers all expect him to fail on the basis of his genetic profile, he shows time and again that he is, in fact, a winner—for instance when he defeats his brother twice in a swimming competition. By contrast, swimming athlete Jerome does not live up to his profile. When he 'only' wins silver at the Olympics and not the much-coveted gold, his world collapses and Jerome tries to take his own life. His suicide attempt fails, and as a result he is condemned to life in a wheelchair.

The tension between genetic determinism on the one hand and the influence of environmental factors and chance or fate on the other is a recurring topic in spoken and written discourse on genetics. Does knowledge of our genetic profile offer us the possibility to know our own future, plan our own lives and take matters into our own hands? Is knowledge indeed power, as Francis Bacon, the famous British philosopher, tells us? Or is it an illusion to think that we will really be able to know our future and plan our lives accordingly? Might it be a better idea to follow the motto of John Lennon and realise that 'life is what happens to you while you're busy making other plans?'

How realistic is it to believe that science fiction stories such as *Gattaca* can become reality? In light of recent technological developments in genotyping and the resulting scientific findings, future scenarios depicting genetic enhancement seem to have gained in credibility. In what follows, we will reflect on a number of assumptions in an attempt to answer the following question: will *Gattaca* remain a world of science *fiction*, or will it soon be more appropriate to speak of science *friction*?

First assumption
Human traits are determined by genetic makeup.

Many human characteristics and disorders, including height, eye colour, blood pressure, intelligence and depression, are to a large extent determined by genetic makeup. A high heritability of body height, for example, indicates that height differences between individuals can be largely explained by differences in genetic makeup, and to a lesser extent by differences in environmental factors such as food, financial means or the availability of social support.

In order to establish the heritability of human traits, researchers need data of genetically related family members, such as parents, children, grandparents and grandchildren, as well as uncles, aunts, nephews, nieces and cousins. These data are used to examine whether individuals who share part of their genetic makeup also resemble each other phenotypically—that is, with regard to physical characteristics and behaviour. The ideal research design to answer this question involves twin research. Since identical twins are more or less genetically identical, any differences between a pair of identical twins cannot be ascribed to genetic makeup but are caused by environmental differences. When a characteristic is very similar in identical twins but not in fraternal twins—who share half their genetic material on average—this is a strong indication that the characteristic is hereditary.

Over the past decades, twin research has provided an important contribution to the mapping of the heritability of various human traits. For example, we now know that body height has a heritability of 80 to 90 % and that psychiatric disorders such as autism and schizophrenia are hereditary to a large extent. This high heritability means that differences between individuals with respect to height, autism or schizophrenia can largely be attributed to differences in genetic makeup. On the other hand, it also means that—in the case of body height—10 to 20 % of the individual differences can be ascribed to environmental factors.

Thus, human traits are only partly determined by genetic makeup. But this is by no means the whole story, as even identical twins show significant differences in personality, intelligence and susceptibility to diseases or disorders. Precisely what determines who we are and how our personalities and bodies develop remains mostly unknown. This is not only determined by our genes but also by our life experiences, what we eat, what kind of substances we come into contact with, and so on and so forth.

Second assumption
Hereditary disorders or characteristics are caused by a single gene.

So-called 'simple' or monogenic disorders are disorders that are caused by one single gene. This is seldom the case, however, as monogenic disorders such as Huntington's disease and cystic fibrosis are rare. It is relatively easy to test for such monogenetically determined disorders.

Most disorders, however, as well as most human characteristics, are of a more complex nature and are caused by multiple genes: they are polygenic. Each gene contributes a small effect, and there is often interaction between genes, genetic variants and environmental influences—in which case we speak of multifactorial disorders. Genetic tests that can establish the presence of one gene have hardly any predictive value in the case of complex disorders or characteristics, rendering them little efficient in diagnostics or prediction. Most geneticists expect that even in the future genetic tests will not be able to predict complex, multifactorial genetic disorders or characteristics. We do know that height is 80 % genetically determined, but we are not even close to knowing which DNA segments and processes are responsible for this.

Nowadays, scientists assume that complex traits are determined by many different genes. In addition, there are indications that the genetic components of complex traits strongly interact with environmental factors. Based solely on genetic makeup, it is impossible to predict the manifestation of a characteristic that also depends on the environment. Therefore, it is realistic to question whether it will ever be possible to select embryos or babies on the basis of a genetic profile that successfully predicts complex traits such as intelligence, height, physical attractiveness or susceptibility to disease—as happens in *Gattaca*.

Third assumption
If a person is genetically susceptible to a condition that is highly heritable, there is no way to overcome or change this.

Individuals who smoke, drink or, for example, gamble often make excuses such as, 'I want to quit, but it is in my genes: I cannot help it'. It is true that the extent to which humans are prone to addictions to for example nicotine is to a large extent genetically determined. This does not mean, however, that they are incapable of giving up smoking (or drinking or gambling).

In the late 1990s, British researchers found that individuals who had been told they were genetically susceptible to cardiovascular diseases—and could therefore benefit substantially from giving up smoking—instead adopted a fatalistic attitude. They believed that their behaviour did not matter anymore because 'it was in their genes', and they smoked more than ever. While the doctors had intended to warn them and point out the benefits of preventive measures (giving up smoking), those concerned interpreted the message as definitive bad news. This reaction can also be seen in *Gattaca* when Vincent's risk of developing a cardiac abnormality is not seen as an increased likelihood for developing a heart disease but as a certainty.

The *risk* of developing a disorder as a result of genetic predisposition is often mistaken for the *certainty* that it will eventually manifest itself. However, risks only express a likelihood, which can, in many cases, be influenced by adjustments in lifestyle or behaviour (an important environmental factor). Discipline and perseverance play an important role. One of the best-known examples in this respect is the disease phenylketonuria (PKU), a heritable metabolic disease characterised by the absence of the enzyme responsible for breaking down phenylalanine. This causes phenylalanine to accumulate in the body, damaging nerve cells and eventually leading to brain damage, cognitive dysfunctions and skin and behavioural problems. PKU is caused by a defect in a single gene, which is responsible for the production of the enzyme that breaks down phenylalanine. This genetic defect can be detected with a simple genetic test. Routine PKU screening has been performed on newborns in many West European countries since the end of the 1960s.

Although the cause of this disease is entirely genetic, whether or not it manifests itself can be strongly influenced by behaviour. By following a strict diet that is low in protein, the accumulation of phenylalanine can be prevented. Even when disease symptoms eventually do appear, possible brain damage can be limited to a minimum by means of a low-protein diet.

Thus, a person who is genetically susceptible to a certain condition has a higher *risk* but in most cases no *guarantee* that this defect or disorder will in fact develop. If and when a disease manifests itself is partly dependent on environmental factors and sometimes even on the interaction between genetic predisposition and the environment.

Fourth assumption
A society based on genetic selection is an ideal society.

In *Gattaca*, the genetic makeup of future generations is determined by means of genetic selection: the optimal combination of genes from both parents. This may seem like a scenario that belongs to a not-yet-realised future, but it is more realistic than it appears at first glance.

For a long time, there has been a taboo on genetic selection because it is reminiscent of Nazi practices during World War II and Adolf Hitler's eugenic programme. In these historical examples, selection was compulsory and there was no freedom of choice for the persons involved. Nowadays, individuals and future parents are allowed to make their own choices when it comes to genetic selection. In Western Europe and the United States, genetic selection against serious and untreatable disorders has gradually become accepted practice. Prenatal screening for Down's Syndrome or *spina bifida* is performed on a large scale, and the technological possibilities in this area are increasing rapidly. As a result, the boundaries of what is considered socially acceptable are changing, along with the routine screening programme offered to pregnant women (see Chap. 11, this volume).

An interesting example, although by no means part of the routine screening programme, is the debate about preimplantation genetic diagnostics (PGD). PGD has been used for several years to screen for a limited number of very severe, untreatable disorders within families with a high risk of developing those disorders. By using IVF, a number of egg and sperm cells are fused outside the body in order to create several embryos. When the cells have multiplied by division, one cell is removed for genetic testing, and only healthy embryos are implanted back into the uterus.

In the spring of 2008, a fierce debate arose in the Dutch media and government circles about whether PGD should also be used to screen for hereditary forms of breast cancer. The discussion focused on three main points: (1) it is not certain that women with an increased risk of breast cancer will develop the disease; this risk varies from 40 to 80 %; (2) hereditary breast cancer can be considered treatable, as women have the option to have both breasts removed as a preventative measure; (3) the disease does not manifest itself until later in life, allowing women to live in good health for many years. Should it be possible to select against the risk of this disease using PGD?

Ten years earlier, at the end of the 1990s, when it had only just been discovered which genes are responsible for hereditary breast cancer, there was a widespread consensus among geneticists and other professionals involved in genetic counselling that prenatal genetic selection should only be used in exceptional cases. In the PGD debate that ensued in 2008, however, the majority of professionals, the public and families concerned had no problems with prenatal genetic selection and even argued that the parties involved had a right to be offered this option. Opposition to this position was restricted to a minority of Christian fundamentalist organisations and individuals.

This gives rise to the question of whether there are boundaries to the development of technological possibilities. Will we in the future also offer genetic tests for intelligence, as happens in *Gattaca*? It is commonly believed that the identification of genes that determine our intelligence could lead to in vitro or prenatal screening of embryos and therefore to genetic selection. We do not think this will ever be possible, as intelligence is determined by many different genes as well as by environmental factors. Even more important, however, is the question of whether high intelligence is more desirable than average or low intelligence. We usually associate intelligence with a greater chance of success and, consequently, happiness, but are persons with high intelligence really happier? This does not always prove to be the case. Research shows that intelligent people are just as happy—or unhappy—as their less intelligent counterparts. Moreover, unlike diseases or disorders, intelligence is a relative, competitive characteristic: if everyone were to become more intelligent, this would not increase our chances of social success.

Let us suppose that we could all agree upon the most desirable characteristics and, by means of genetic manipulation (switching genes on or off), were able to ensure that the next generation would never be ill and would be exceptionally intelligent, handsome, fit, patient and empathic. In that case, there would be little variation in these characteristics—after all, everyone would be genetically selected, or even manipulated, for these traits. Society would become rather dull! In the space station in *Gattaca* we already encounter such uniformity, as the future astronauts show little diversity in appearance, intelligence and behaviour.

A far greater problem, however, is that this lack of variety would cause everyone to have the same ambitions and abilities. If everyone wanted to become the manager of a large company, there would be no one left to work under their management. In that case, we would have to be selective and turn on the genes for good leadership in some individuals (managers) and those for taking orders in others (employees). But who decides on what grounds embryos will be preselected for a particular position in an organisation (or society)?

With this, we touch upon one of the greatest fears associated with eugenic programmes: who will have access to the possibilities of genetic selection and manipulation? As the procedures are likely to be costly, we might reasonably fear that only the wealthy will be able to afford them.

Will Genetic Enhancement of Human Beings be Possible in the Future?

In the previous sections, we have refuted all four assumptions. The short answer to the question of whether genetic enhancement of humans will be possible in the future is therefore 'no'. However, this conclusion requires further nuance.

As has been mentioned previously, there are a few exceptional situations that do allow for genetic enhancement. When a disease is 100 % genetically determined and is entirely influenced by a single gene, it is possible to determine whether someone will become ill or not by using a genetic test. This will allow them to take appropriate precautions concerning lifestyle or therapy, which is usually much easier than genetic manipulation. Similarly, when a characteristic is caused entirely by environmental factors and by one factor in particular (for instance, being exposed to radioactivity in the case of cancer) it is advisable to avoid that factor.

In sum, both genetic and environmental factors contribute to human traits and their possibilities for enhancement. Before we can change them, however, these factors and the way they are interrelated (and interact) should first be established. Since most characteristics are not determined by a single genetic or environmental factor but by many, it is not likely that human traits can be enhanced on the basis of genetic manipulation or screening.

Genetic human enhancement is ultimately limited because of the complex relationship between genes and environment. Humans are complex beings, also from a genetic point of view. On the one hand, they are the product of their genes, but on the other they are greatly influenced by the environment in which they spend their lives, from conception to death. Human beings are formed by the presence or absence of incentives and stimuli in this environment. The interaction between genetic predisposition and environmental factors cannot be controlled and is to a large extent unpredictable. Especially for traits such as intelligence—which is caused by a complex interaction between multiple genes and the environment—it is not likely that there will ever be a predictive prenatal test.

Conclusion

The topic of this book is human engineering or human enhancement, a phenomenon that increasingly touches upon the reality of the twenty-first century. The field of genetics has made no small contribution to this advancement, for example by the development of genetic screening tests for rare, monogenic disorders. The application of genetic screening or even genetic manipulation for characteristics such as personality, intelligence or appearance still remains impossible—a situation that is not likely to change soon. The possible enhancement of these characteristics is to a large extent determined by environment and personality, rather than genetic manipulation. And this brings us back to *Gattaca*: genetic makeup indicates only chance or capacity. Whether or not a characteristic manifests itself depends on many additional factors, including perseverance, personality, ill fortune and sheer dumb luck.

Bibliography

Deary IJ, Johnson W, Houlihan LM (2009) Genetic foundations of human intelligence. Hum Genet 126(1):215–232 *Provides a good overview of the heritability of human intelligence*

Galton F (1865) Hereditary talent and character. Macmillan's Magazine 12:157–166, 318–327 *Sir Francis Galton was the first to systematically describe ideas about eugenics in this classic paper*

Gonzales JL (2011) Ethics for the pediatrician: genetic testing and newborn screening. Pediatr Rev 32(11):490–493 *A nice overview of ethical issues in newborn screening*

Chapter 6
Gulliver's Next Travels: A Journey into the Land of Biomaterials and Synthetic Life

Annemiek Nelis and Hub Zwart

Abstract This chapter explores the state-of-the-art in research into the synthesis of new biomaterials. The underlying question is to what extent it may become possible in the future to create new life or possibly even human life using these materials. Recent research in this area has produced materials such as a synthetic equivalent of spider silk, which is biodegradable, can be broken down by the human body and is therefore ideal for medical applications. Another possible application of such biomaterials is 'smart drug delivery', a technique in which nanocapsules transport drugs or genes to specific parts of the body and which could, for example, be used to treat metabolic diseases. When made of biosynthetic materials, these capsules could be broken down by the body, thus minimising the chance of side effects or rejection. Eventually, scientists hope to be able to create new life using these biomaterials.

Lemuel Gulliver's journeys to the lands of the Lilliputians and the giants, as described in *Gulliver's Travels*, are known all over the world. Less well-known, however, are the book's two other stories, which recount Gulliver's travels to the countries of the scientists and horses. In 'Voyage to Laputa', Jonathan Swift

---

Translated by Lydia ten Brummelhuis.

A. Nelis (✉)
Dutch Safety Board, P.O. Box 95404, 2509 CK, The Hague, The Netherlands
e-mail: a.nelis@safetyboard.nl
URL: www.safetyboard.nl

H. Zwart
ISIS—Department of Philosophy and Science Studies, Radboud University Nijmegen,
P.O. Box 9010, 6500 GL, Nijmegen, The Netherlands
e-mail: h.zwart@science.ru.nl

B.-J. Koops et al. (eds.), *Engineering the Human*, DOI: 10.1007/978-3-642-35096-2_6, 71
© Springer-Verlag Berlin Heidelberg 2013

(1667–1745) describes Gulliver's visit to the Academy of Lagado on the scientist island of Laputa.

'Voyage to Laputa' is a parody of *The New Atlantis*, a greatly influential book written by Francis Bacon (1561–1626), Lord Chancellor of England. Being a defence of large-scale scientific research financed by the government, this work depicts a scientific research institute that is situated on an unknown island, the new Atlantis. This is an ideal community of scientists—a kind of scientific monastery or academic utopia.

Written in the guise of a travel account, Bacon's book contains important proposals for the organisation and financing of scientific research, making the government an active partner in both. The research institute where the researchers are stationed, Salomon's House, is described in detail. The scientists are called 'fellows', and the work has been efficiently distributed: 'Merchants' collect books and instruments, 'Pioneers' design new experiments, 'Inoculators' conduct the experiments, 'Compilers' present the results of these experiments in tables, 'Interpreters' interpret the results in terms of 'aphorisms' and 'Benefactors' contemplate their practical application in society.

The New Atlantis was highly influential in the founding of the Royal Society in London, whose members are also called 'Fellows'. It is precisely this Society, involved at the time in laying the foundation for the experimental method in the natural sciences, which Swift ridicules in his 'Voyage to Laputa'. Swift criticises the manner of conducting science that Bacon advocates and particularly targets the results of empirical research, questioning both their relevance and usefulness. He describes meetings and conversations with unkempt researchers or 'projectors' who are out of touch with reality and work on ridiculous questions and projects in an immense laboratory complex containing over 500 offices. For the naive visitor—Lemuel Gulliver—their research seems to lack all practical relevance. The researchers demonstrate their 'contrivances' or technical instruments and assure visitors that they are about to make world-shocking discoveries—provided that the outside world is prepared to supply extra financial resources.

The approach advocated by Bacon—the large-scale financing of scientific research by the state—was ridiculed by Swift as the financing of hobbies of other-worldly scientists. Nowadays, however, public funding of science is hardly controversial, as science is expected to contribute to important economic and social developments. The distinction between fundamental and applied research has become less and less relevant. According to Helga Nowotny, contemporary scientific research is nearly always 'applied' and generally takes place in what she calls 'the context of use'; scientific issues are usually related to societal applications, and, from an early stage on, they are formulated in dialogue with possible users.

One aspect of contemporary research that has not changed compared to the world described by Gulliver—and to a lesser extent Bacon—is its large scale. Scientific research is increasingly conducted by collaborative research consortia that comprise several research teams, organisations—that is, universities and companies—and countries.

6 Gulliver's Next Travels: A Journey into the Land of Biomaterials and Synthetic Life   73

This contribution will discuss research on biomaterials (one of the strands of research ridiculed by Gulliver!) as an example of large-scale programmes in which fundamental questions are explicitly linked to societal applications. The main focus will be on new biomaterials—materials made from vegetable or microbial sources—and the question of whether in the future it will be possible to generate new life from these materials. In the field of biomaterials, molecular knowledge and techniques are used increasingly often. The chapter is based on a number of conversations with Jan van Hest, who is professor of Bioorganic Chemistry at Radboud University Nijmegen and who was a member of the Young Academy of the Royal Netherlands Academy of Arts and Sciences at the time.

Biomaterials

Van Hest's research focuses on the production of new materials in general and on biomaterials, which are created through a combination of biology and chemistry, in particular. So far, new materials have mainly been produced from raw materials such as natural gas and oil. Plastics are a good example. In general, however, plastics and other synthetic materials are not biodegradable, and although they can be highly useful for many purposes, they do not compare with materials produced by nature—the latter being not only biodegradable, but often also strong, flexible, supple, durable and self-restoring. For all their sophisticated assembly techniques, human beings still cannot compete with nature in the creation of biomaterials.

Van Hest and his colleagues draw their inspiration directly from nature, as has become characteristic of the field of biosynthetic materials. Increasingly often, microorganisms are called upon for assistance, and recombinant DNA techniques have enabled the synthetic production of natural substances such as human insulin or human growth hormone in laboratories. Already, microorganisms are used in the production of insulin, but in theory it should be possible to artificially prepare a wide variety of biologically active substances, including protein-based medication.

After the DNA sequence—and therefore the constituting amino acids—of a substance such as growth hormone has been established, this sequence is replicated in a lab and subsequently used to produce synthetic strands of DNA. Small changes made to the synthetic DNA compared to the original DNA ('point mutations') render it possible to create new protein molecules in microorganisms. This method is used, for example, in the production of pharmaceuticals.

In the examples of insulin and growth hormone, nature is *imitated*. Initially, molecular biology was mainly applied to the reproduction of proteins, but in recent years scientists have come to realise that many natural materials consist of proteins, and can therefore be created and manipulated in the lab by means of molecular techniques. In addition, nature can also serve as a source of inspiration for the development of *new* materials based on the existing ones. The biosynthetic material van Hest is working on offers the possibility to create complex and refined products in this manner, especially on the micro and nano scale.

One of van Hest's main research topics is spider silk. As early as the eighteenth century, Gulliver already realised that spider silk would make an ideal model organism. While roaming the laboratory complex on the island of Laputa, in one of the many chambers Gulliver meets someone who researches the structure of spider webs.

> I went into another room, where the walls and ceiling were all hung round with cobwebs, except a narrow passage for the artist to go in and out. At my entrance he called aloud to me not to disturb his webs. He lamented the fatal mistake the world had been so long in of using silkworms, while we had such plenty of domestic insects, who infinitely excelled the former, because they understood how to weave as well as spin. And he proposed farther that by employing spiders the charge of dyeing silks would be wholly saved, whereof I was fully convinced when he showed me a vast number of flies most beautifully coloured, wherewith he fed his spiders... [he hoped to] find proper food for the flies, of certain gums, oils, and other glutinous matter to give a strength and consistence to the threads. (Swift 1726/1967, p. 225).

Spider silk, Gulliver is told by the researcher, is a unique material: light, pliable and strong. Thus far, however, we have failed to make use of such materials. In his experiment, the scientist tries to indirectly modify the characteristics of spider silk, starting with its colour. Since spiders catch flies and suck their juices, he offers insects of different colours to a group of spiders in order to determine whether it is possible to influence the colour of their silk.

Swift uses research into spider silk to illustrate the absurd character of the scientific experiments conducted at the academy of Laputa. Nowadays, however, the project is no longer absurd and has even become reality. It has quickly developed from an object of ridicule into one of the most important areas of contemporary research, both from a scientific and societal point of view. Just like the researcher in Laputa, van Hest emphasises the unique character of spider silk, claiming that we have seriously neglected this material in the past and literally cast it aside as waste—erroneously so. In fact, he argues, we should be jealous of spiders, as they are able to produce a material that we cannot or barely imitate, but could very well use. Spider silk is both strong and tough—it cannot be easily broken nor does it bounce in all directions. These qualities make spider silk unique.

Compared to natural materials such as spider silk, or more broadly, protein-like fibres in general, human-made artificial materials such as plastics are highly primitive—that is, on the micro level. The microorganisation of the synthetic materials we have developed in the past is relatively simple. The organisation of proteins, however, is far more complex, and further insight into their structure would open up many possibilities for developing new materials with special features or for improving the existing ones. Another important characteristic of biomaterials is their biodegradability, which is not only relevant for reasons of environmental hygiene, but may also be important for biomedical applications. Biodegradable suture thread, for example, is more user friendly than synthetic material, as the former can remain in the body while the latter has to be removed again.

In recent years, the biological structure of spider silk has been closely examined. Van Hest and his colleagues are interested in the production of 'biosynthetic' spider silk, a new material that resembles naturally occurring spider silk but is manufactured by means of biological and chemical techniques. Biosynthetic spider silk would be a simplified but technically reproducible substitute of natural spider silk. The secret of biomaterials is their arrangement on the molecular level. In this particular respect, contemporary researchers wish to follow nature more closely, for example by using—instead of spiders—*E. coli* bacteria that can manufacture new conformations of amino acids. This technique generates new synthetic polymers which, compared to those produced by traditional methods of synthesis, are uniquely arranged on the nano level.

According to van Hest, it would be going too far to suggest we could match nature by producing materials identical to naturally occurring ones. Rather, scientists intend to learn from nature by producing 'analogous' or at least biocompatible materials, for example by the partial imitation of natural processes; they draw out the essence, and replace those aspects they do not yet understand by synthetic materials. Current research focuses on the development of a material which displays several of the properties of spider silk. Even that would be an important step forward in our search for products that are stronger, more flexible and less damaging to nature— that is, biodegradable. Researchers have discovered, for example, that spider silk consists of molecular-sized strings. For biomedical applications it is important to gain more control over the materials' microorganisation. By developing materials to which specific proteins attach, for instance, it becomes possible to produce applications such as biosensors. The proteins in spider silk are stacked consecutively as small plates, allowing for both conduction and crystallisation. Van Hest's team focuses on two proteins: a protein that is responsible for the *elasticity* of tissues and a protein that is responsible for the *strength* of the silk and prevents it from breaking easily. Both proteins have a relatively simple structure.

Van Hest expects to find potential applications for synthetic spider silk and other biomaterials mainly in the biomedical domain, seeing that in the medical context a material's effectiveness is more important than its price. A possible application involves spider silk-like tissues that could in the future be used by surgeons to close wounds or heal burns without leaving scars. This application is an important challenge for this type of research. Another possibility lies in tissue engineering, a technique that would enable us, for example, to close the palate without scarring in the case of schisis (harelip or cleft palate), or, in the case of congenital heart diseases, to cultivate a new heart valve.

Nanofactories in the Cell

Another possible application of contemporary materials research is 'smart drug delivery'. This technique employs capsules that were developed using nanotechnology—that is, on a supramolecular scale—and which 'transport' substances through the human body. Thus, medication (protein therapy) and genes (gene therapy) can be conveyed to specific places in the body. Van Hest mainly focuses on protein therapy. Already, liposomes—hollow fat globules—are often used as a means of transportation for the delivery of medication, for example in the treatment of cancer. In order to ensure the immune system does not mistake certain molecules for enemy intruders—and consequently rejects them and tries to rid itself of them—they are enveloped in a biosynthetic capsule, a process called 'cloaking'. This capsule is covered in molecules that seek out particular cells (the so-called 'targeting device'), and not until they have found these cells is their content released and are the molecules activated. The ultimate goal is to produce these capsules from natural or biosynthetic materials which can be broken down by the body, thus minimising the risk of side effects or rejection.

Van Hest and his colleagues are working on the production of nanocapsules, loading them with various enzymes in order to provoke a number of reactions. They are interested in getting these capsules inside a cell and creating an artificial organelle. The human cell can be regarded as a complicated factory in which enzymes (the 'workers' of natural synthesis) incessantly produce a wide variety of biomaterials from raw materials such as amino acids, sugars, nucleic acids and lipids. The creation of an artificial organelle concerns only part of the cell. A cell contains a large number of 'rooms' where different processes take place: there are rooms that produce energy (the mitochondria); there is a room that stores the genetic information (the nucleus); and there are rooms that break down excess proteins and clean up the waste. These last rooms need to be well separated from the other sections of the house—the rest of the cell—in order to avoid disruptive effects such as the breakdown of the entire cell. By supplying one specific room with proteins, it becomes possible to add new functions to a cell or to restore the existing ones. Van Hest and his team have created a capsule that is similar in size to a single room or organelle and that was designed in such a way as to enable them to encapsulate something in the organelle, while at the same time allowing the capsule to communicate with the other rooms. This way, small molecules can enter and leave.

Van Hest hopes his research into nanocapsules will contribute, among other things, to the development of a therapy for metabolic diseases. There are hundreds of known metabolic diseases, often caused by the insufficient breakdown of a single substance in the body. As a result, an excess of this substance accumulates, leading to various harmful effects. In many cases, only one enzyme is needed to break down the substance, and thus remedy often severe symptoms.

Synthetically produced nanocapsules are not yet suitable for clinical application. One of the problems that remains to be solved concerns regulation: how does one ensure that neither too much nor too little is broken down? According to van Hest, it is essential to develop a regulation mechanism; while we might be able to convert certain substances, without a braking mechanism the consequences can be serious. In addition, proteins are swiftly broken down in a cell and therefore quickly lose their function. Unlike gene therapy—which ultimately aims to change genes permanently—protein therapy still has a short-lived effect. By using capsules, van Hest and his colleagues hope to maintain this effect for a longer period of time.

Creating Life

As mentioned above, van Hest and his colleagues are working on a synthetic organelle. Eventually, they hope to extend their research to other parts of the cell as well, imitating the separate rooms. In the future they may be able to group those different rooms together, thus creating a larger whole and possibly even a functioning system. Ideally, the chamber wall or cell wall, which is still produced synthetically, will soon also be made from natural components. Then it might become possible for the room to replicate itself, rendering the treatment of metabolic diseases more natural and its effects longer lasting.

Whether biomaterials will ever be able to reproduce remains to be seen. This question is related to another question that has long preoccupied scientists: is it possible to create synthetic life? While only 10 years ago many scientists still considered this to be impossible, its realisation now seems to be getting nearer. In 2008, American researchers for the first time created a synthetic cell, which, although still rather primitive, in essence resembles nature. It has a cell wall that contains genetic information, this information is read and subsequently copied, and afterwards the cell divides into two parts, producing two separate capsules. This already approaches a self-replicating system and therefore synthetic life.

In addition to the chemists' approach, which focuses on the creation of a synthetic cell, there is a second route towards developing synthetic life: that of molecular scientists. This approach constitutes the fabrication of a synthetic genome—the entirety of an organism's DNA. An important figure in this field is Craig Venter, a researcher who played a major role in the deciphering of human DNA and who is trying to establish the minimum number of genes necessary for a bacterium to survive, stripping bacterial cells of all genes they do not necessarily need. The resulting 'minimal genome' could be the basis or chassis for the construction of new cells, where it functions as a host system into which an artificial genome is implemented. This synthetic genome is subsequently used by the host system to produce new daughter cells containing a copy of the synthetic genome. The thought behind this approach is that the chassis can be used to build genetic networks which can in turn be used to produce new, customised genetic combinations. Venter's ultimate goal is to build so-called 'bio-bricks'—comparable to Lego bricks—in order to make proteins and molecules that do not naturally occur in cells.

Van Hest believes that Venter's method, rather than the chemists' approach, has the greatest chance of being the first to succeed in generating new life. This is due, among other things, to the fact that Venter stays close to the natural processes and uses the existing building blocks, whereas chemists are trying to create bio-building blocks—that is, living building blocks—from non-natural components. In other words, Venter is using a computer that is ready for use; all that remains to be done is to insert a chip and an Intel-processor in order to boot the system. The chemists, on the other hand, are using various separate components to build a computer from scratch.

The huge significance of synthetic life indicates a substantial change in our view of life. If we succeed in creating synthetic life from previously non-living materials, this will be a breakthrough comparable to the work of Friedrich Wöhler in the nineteenth century. Wöhler was the first to succeed in synthesising an organic molecule from two *non*-organic compounds, a feat formerly considered impossible as organic molecules could only be isolated from nature. The molecule he created was urea, a waste product formed by the breakdown of proteins and discharged by the kidneys. The synthetic form Wöhler produced is used as a plant fertiliser. While nature seemed to hold the monopoly on the creation of molecules, humans now turned out to be capable of the same feat. Wöhler's work caused a paradigm shift in science—a radical break with contemporary thinking.

A comparable shift in paradigm was brought about by Darwin, whose works caused human beings to lose their separate status in biology and to become part of the system of living beings. Such discoveries drastically change our worldview, as would also be the case with synthetic life. If scientists are indeed capable of creating artificial life, this would essentially break down several religious and philosophical barriers, the principle of making life having so far been reserved for nature.

Whether there will indeed be a paradigm shift remains to be seen, but expectations are running high. In the meantime, an additional question arises: what exactly constitutes life? One way of defining life is to look at the extent to which an organism—or a collection of molecules—is able to perform the basic function of life: to sustain itself. This comprises the consumption of energy and food on the one hand and reproduction on the other. In other words, life ensures its own continued existence.

According to van Hest, the creation of a synthetic cell and the rise of synthetic biology have sharpened the definition of 'life'. For, in addition to energy, nutrition and reproduction, one could argue that some kind of evolutionary process should also be a criterion for life; a system should not be considered new life until it continues to evolve and, instead of merely dividing, also adjusts to its environment. For example, Venter has already succeeded in producing molecules and proteins that do not naturally occur in cells, but does this make them a form of new life? Not really, because Venter uses existing molecules that, although cleverly manipulated, are not yet capable of surviving independently as a system, let alone adapt evolutionarily to new conditions.

This is not to say, however, that Venter will not develop a living system that does meet these requirements in the foreseeable future. Van Hest estimates that within the next 20 years somebody on our planet will be able to create synthetic life, thus claiming the right for humans to create life from originally non-living components. Since we generally attribute life to the soul or God, such a discovery would cause a considerable shock and call our beliefs into question, effecting a radical shift in our ways of thinking.

Creating Humans

If in the future it will be technically possible to create new life, will it then also be possible to create human beings? Van Hest is sceptical. We might be able to make single-celled organisms or a new microorganism, but humans are far more complex. The information or software that is present in human beings—or in any multicellular organism, for that matter—is too complex for us to understand, and van Hest doubts whether we ever will. But then, never say never.

Venter's project is only the beginning. In essence, he replaces one element of an existing life form by introducing an artificial genome. This only changes the genome, however, which is still far removed from the complexity of an entire organism such as a human being. In effect, van Hest claims, Venter uses the cell as a host into which he only needs to plug a software programme. All that can be done, therefore, is to make the genome as small as possible in order to reduce its complexity—Venter's chassis or minimal genome. In addition, we can use the existing genomes, which are combined by scientists in various ways. Cloning is an important technique in this respect, and one that—for Dolly the sheep at least—has proven its effectiveness.

According to George Whitesides, the world's most-cited chemist, it is difficult to establish what precisely constitutes life. Since life is such a complex and unintelligible process, he considers it impossible to imitate. Van Hest agrees with Whitesides' point about humans' complexity, but still believes it possible to increase our chances of success by choosing clever structures, taking smart examples and closely watching nature.

Where human beings are concerned, complexity is here to stay. In order to understand complex systems—that is, systems so complicated we do not really understand what they are—we should try to isolate laws. For structures as complex as life or living human beings, however, it does not suffice to look only from the perspective of biology or the molecular sciences. Mathematicians, physicists and informaticians are also needed: scientists who can define laws, formulate models and help organise the mess of data. Only through cooperation can the complexity become comprehensible, so that after an inventory of all the different branches, it will be possible to disregard most of them for the sake of simplicity. Thus, only the core of the system will remain. For the time being, however, van Hest lacks the necessary overview and is still unable to see the wood for the trees.

In short, the complexity of life not only calls for new knowledge and techniques, but also for new forms of cooperation. To tackle this issue, van Hest argues, we need to strive towards a higher level of multidisciplinarity, in which the mathematicians, informaticians and physicists mentioned above are of crucial importance. Walls between disciplines are being torn down—not only metaphorically but also literally in the physical design of modern laboratories. Researchers cooperate more and more, and various disciplines are brought together in research teams and institutions, including cellular biology, organic and inorganic chemistry, physics, information technology and the biomedical sciences. The new generation of scientists, according to van Hest, is interested in the meeting point between chemistry, nanotechnology and biotechnology.

This kind of cooperation is already taking place at a number of locations in the world. At first it generates a lot of noise, as scientists from different disciplines do not fully understand each other. But eventually, van Hest claims, the exchange between disciplines and disciplinary insights will be crucial in understanding the high degree of human complexity. What order is there in the current disorder? Can we create order in this complexity? At the moment, scientists are handling these questions rather intuitively; many of the experiments that are currently being conducted were shaped by intuition and involve a lot of tinkering at the margins of life, aiming to better understand its complexity. Van Hest does not know if—or how far—we will succeed in this undertaking.

Conclusion

What would happen if the author of *Gulliver's Travels*, Jonathan Swift, were to write a second book about the development of science today? Surely, he would not use the manipulation of spider silk as an example to depict an idiosyncratic scientist who is only interested in his own hobbies, for nowadays spider silk is studied with enthusiasm and success by biochemists such as Jan van Hest. Swift might get upset, however, by the large sums of public funds spent on scientific research. On the whole, contemporary science matches the Baconian ideal: it is heavily reliant on public financing and conducted increasingly often in the service of social and economic developments. Research into fundamental questions—such as, in van Hest's case, whether it is possible to create synthetic life—is closely connected to medical and societal applications, including medication, suture materials and genetic tests. In this respect, contemporary research leaves less scope for ridicule and satire.

An aspect of modern science that might become a target of Swift's satire is the rapid pace at which new developments occur. The scientific context in which Jan van Hest operates is exemplary of this speed. While in our first conversation he mentioned only a few scientists who dared to speculate about the possibility of producing synthetic life, in a second conversation, barely 18 months later, van Hest himself expressed more confidence in the project's chances of success. The molecular life sciences are characterised by a relatively recent and turbulent history. In 1953, the structure of DNA was elucidated by Cambridge scientists James Watson and Francis Crick, and since then there have been great changes both in the understanding and manipulation of life. The human genome has been charted, DNA has been produced in synthetic form, and who knows, one day it may even be possible to create life from non-living materials.

At the same time, we saw that scientists are only beginning to understand life, which turns out to be complex—both at the cellular and human level. What does the combination of the sometimes frightening speed at which science is developing on the one hand and the complexity of cells and organisms on the other mean for the future of the human race? We do not know. It is a question that Lemuel Gulliver may want to ask on his island of the future. Is it not time for Gulliver to travel again?

Bibliography

de Vriend H, van Est R, Walhout B (2007) Leven maken. Maatschappelijke reflectie op de opkomst van synthetische biologie. Rathenau Institute. Working document 98, June 2007

Lensen D, Vriezema DM, van Hest JCM (2008) Polymeric microcapsules for synthetic applications. Macromol Biosci 991–1005

Nowotny H (2003) Democratising expertise and socially robust knowledge. Sci Publ Policy 30(3):151–156

Swift J (1726/1967) Voyage to Laputa. In: Gulliver's travels. Penguin, Harmondsworth

van Hest, JCM (2001) Natuurlijke inspiratie voor een geïntegreerde wetenschap. Inaugural lecture, Radboud University Nijmegen, Nijmegen

van Hest JCM, Tirrell DA (2001) Protein-based materials, toward a new level of structural control. Chem Commun 1897–1904

Venter JC (2007) A life decoded—my genome, my life. Viking/Penguin Group, New York

Chapter 7
Human Robots and Robotic Humans

Catholijn M. Jonker and Annemiek Nelis

Abstract Humans and machines are, at present, viewed as two distinct categories differentiated by their unique biological or artificial configuration and capabilities. This chapter argues that the two will, in the near future, increasingly partake of the nature of the other as a result of technological enhancement. A number of recent developments within the field of robotics are already causing the traditional boundaries between humans and machines to fade, both as a result of mankind's increasing reliance on modern technology and of technological developments which allow modern machines to imitate human behaviour to a high degree. Modern technology thus has the potential to aid and even modify the human organism by artificial means. The construction of such advanced forms of humanoid machines, or 'androids', requires a reconceptualisation of what makes humans characteristically human, as well as requiring us to pay attention to the potential consequences of constructing such humanlike machines.

Translated by Geert de Jong.

C. M. Jonker (✉)
Section Interactive Intelligence, Department of Intelligent Systems,
Delft University of Technology, Mekelweg 4, 2628 CD, Delft, The Netherlands
e-mail: c.m.jonker@tudelft.nl

A. Nelis
Dutch Safety Board, P.O. Box 95404, 2509 CK, The Hague, The Netherlands
e-mail: a.nelis@safetyboard.nl

B.-J. Koops et al. (eds.), *Engineering the Human*, DOI: 10.1007/978-3-642-35096-2_7,
© Springer-Verlag Berlin Heidelberg 2013

> It has been said that most people overestimate how much technological progress there will be in the short term and underestimate how much there will be in the long term. (Bostrom 2006, p. 47)

In 2008, the Dutch University of Maastricht hosted an international conference on the theme of 'Human–Robot Relationships'. The conference was convened to assess the possibility of humans and robots forming intimate, loving relationships in the future. This topic proved a source of considerable disagreement among the 30 participants. Not only were they divided on the question of how far technological advancements could take us, but many were also undecided as to whether there was any need for such personalised relationships with robots in the first place. 'If intimate relationships with robots is the answer,' one participant went on to remark in the Dutch newspaper *NRC Handelsblad* (issue of June 21, 2008), 'then *what on earth* is it an answer to?'

Not yet a reality, such relationships are currently only found in novels and films. Consider, for example, the 2001 film *A.I.* (*Artificial Intelligence*) by Stanley Kubrick and Steven Spielberg. In this film a fictional world is depicted in which scientists have managed to create a boy robot called David, who is not only capable of learning and operating autonomously, but who can also experience love. An older example of this genre is the well-known series of robot narratives written by American author Isaac Asimov. A biochemist by training, Asimov wrote nine futuristic stories between 1950 and 1960 in which he envisaged ever-evolving robots that could come to serve as perfect substitutes for actual human beings.

As yet, David and the Asimov robots are confined to the realm of fiction, existing only as imaginary figures on screen or on paper. Scientists, however, do hope to turn fiction into reality, by continuing to develop sophisticated robotic machines that resemble humans in appearance and behaviour and by finding ways of making them interact with their surroundings in an intuitive manner. Such machines are designed to replicate human movement and to mimic and respond to human behaviour just as David and Asimov's robots would. Still, they are not generally considered human: although they share many characteristics with humans, they remain machines at heart. Whoever attempts to cut open David's body will not draw blood from his vessels, but will instead encounter computer chips and an interior riddled with cables and wires.

In our contribution to this volume, we will point to a number of developments emerging with regard to the construction of robots that look and behave more and more like human beings. At the same time, another development can be discerned which could be called the mirror image of this increase in the construction of humanlike androids: namely, our own increasing resemblance to machines through the application of computer technology. As a consequence, *homo sapiens* is slowly becoming a cyborg, a physical merger of man and machine. In popular culture, examples of machine-enhanced beings are already heavily featured, such as *Star Trek's* 'Borg', the race of enhanced humanoids made famous by the popular television series.

The most important thesis of this chapter can be stated as follows: as a result of the ever-growing resemblance between humans and machines, the boundary between what constitutes a human and what constitutes a machine is becoming less distinct. This will have implications for how we conceive of ourselves as human beings and for what it means to live in a 'human-oriented' society.

We first turn our attention to humanoid cyborgs, by discussing a number of cases where computer networks and chip implants have enabled humans to communicate both with one another and with objects such as electronic doors and wheelchairs. Subsequently, we touch on the construction of android robots that are capable of mimicking gestures and movements specific to humans. Lastly, we review several examples of robots which possess a strong visual likeness to their human creators. Included in the reference list at the end of the chapter is a list of websites containing more information on each of these topics. Central to our discussion is a concern with the similarities and differences between humans and machines, leading to the question of whether machines can operate with a degree of autonomy from their human inventors.

Part Human, Part Machine

British scientist Kevin Warwick is renowned for being an extraordinarily creative experimenter. He achieved great successes experimenting not just in a lab using computers, but by conducting experiments on himself as well. In 1998, he had a small chip inserted into his arm which enabled him to maintain a wireless connection to a computer. With this experiment, Warwick aimed to show how computers and humans might communicate without relying on the use of a mouse or a keyboard. Warwick is of the opinion that in the future we will all be hooked up to computer networks with which we share a continuous, wireless rapport.

By means of radio waves, the computer was able to pinpoint Warwick's location. Furthermore, it was programmed to respond to several of his activities:

> At the main entrance, a voice box operated by the computer said 'Hello' when I entered; the computer detected my progress through the building, opening the door to my lab for me as I approached it and switching on the lights. For the nine days the implant was in place, I performed seemingly magical acts simply by walking in a particular direction. (K. Warwick, 'Cyborg 1.0', www.wired.com/wired/archive/8.02/warwick.html, accessed September 2012)

In 2002, another experiment was carried out where the chip implanted into Warwick's arm was not only linked to the external computer network, but was also connected internally to his central nervous system. In this experiment, the computer not only tracked and responded to movement, but it also enabled Warwick to steer a wheelchair and even guide an artificial hand. Thus, he had found a way to both receive and send signals from and to a computer from a distance.

A possible practical application of this latter type of experiment could be to aid patients who are physically impaired as a result of the loss of a limb or who are otherwise restricted in their ability to move around freely. An interesting question raised by this experiment would be whether such biomedical instruments should just be considered as expedients for helping human beings with disabilities, or whether they are part of the development that constitutes the creation of modified or enhanced humans.

Warwick himself is outspoken in his answer to this question. He believes the chip will not only replace existing body functions, but will also give rise to a set of wholly novel applications. Theoretically, those who carry a chip in their arm will be capable of more than immediate, localised operations. They will be able to execute commands from anywhere in the world by virtue of being connected to a worldwide network of intelligent systems.

As an extension to this experiment and in order to demonstrate its effectiveness, Warwick equipped his wife with a simpler version of the chip. By connecting the computers that received the chips' signals, each person was made aware of the other's actions—even when they were at separate locations. Mrs Warwick could sense the actions performed by her husband, such as lifting a finger, and vice versa. The experience of undergoing such an experiment was described by Warwick as getting a sense of what it would be like to be an actual cyborg:

> I was born human. But this was an accident of fate—a condition merely of time and place. I believe it's something we have the power to change. (http://www.wired.com/wired/archive/8.02/warwick.html, accessed September 2012)

BrainGain: A Built-In Computer

The research project set up by the Dutch BrainGain consortium also revolves around the possibility of enabling direct communication between a computer system and the brain. Its primary focus is on improving connections between brain and computer by means of 'Brain Computer Interfacing' (BCI). Users of BCI are trained to concentrate on specific thoughts in order to mentally control a computer or to receive electronic stimuli to the brain. BrainGain is a project involving co-operation from several universities, university medical centres, the Dutch research institute TNO, patient advocacy groups and electronics companies such as Royal Philips and Siemens. These partners share the goal of investigating to what extent individuals can learn to control their brainwave activity by means of neurofeedback training, a process where users consciously activate certain brain regions in order to operate computers by thought alone.

One of the findings of modern neuroscience is that when a person passively observes certain instruments or tools, activity is also triggered in those brain regions that are involved in the actual use of the instruments. Furthermore, it appears that the precise location of the brain activity is determined by what type of tool is being observed. BrainGain investigates whether it can build on these insights. If, for example, the act of just *thinking* about a tool similarly results in discernible patterns of brain activity, it might be possible for a person to instruct a computer to operate the tool in question—provided the brainwave patterns are sufficiently pronounced.

Ideally, this kind of research will bring solutions to the problems of persons who suffer from poor mobility or from a communication deficit. In addition, patients suffering from a form of muscle dysfunction such as ALS, epilepsy or paraplegia are also likely to reap benefits from it. Due to an increased ability to actively interact with their surroundings, these patients will require less care and assistance in performing daily tasks. Training them to focus on a single sound, object or movement might result in the kind of neural activity required to operate a computer, which can be programmed to perform useful functions such as opening doors or switching on the lights.

BrainGain resembles the second Warwick experiment in its ambition to provide disabled patients with technology that will improve their quality of life. A possible advantage comes from the fact that the computer receives signals directly from the brain rather than from a chip embedded in the arm. Once this technology has reached the appropriate stages of development, it will also offer solutions to patients with a damaged connection between brain and body parts such as the arm, for instance as a result of injury or amputation.

A Closer Look at the Work of Warwick and BrainGain

A central objective of both the BrainGain project and Kevin Warwick's research programme is to find ways of compensating for the loss of basic human skills in individuals. A case in point are persons whose freedom of movement or communicative abilities have been curtailed as a result of muscle loss or speech impairment. Researchers devote much effort towards developing medical instruments and accessories that might be of help to these patients in compensating for their disabilities. Thus, one major impetus behind such research is that it offers possible solutions to persons with a work-limiting disability.

Besides *compensating* for disabilities, which is mainly of benefit to elderly people and patients with a chronic illness or physical disability, this technology can also be used to facilitate the *improvement* or *enhancement* of unimpaired human capabilities. The use of such intelligent technology might thus also open the doors to novel applications. In Warwick's case, having a chip in his arm meant he was capable of more than operating doors and switching on lights. The chip also linked him to his wife through a computerised network that allowed him to communicate with her from afar in a new way.

One of the questions raised by such experiments is how far Warwick and his colleague scientists will go in their quest for human improvement. This issue is linked to matters of controlling the new technology. Will man or computer be in charge of the controls for operating this technology and for switching it off? Usually we are led to think that computers are under the direct control of humans. After all, are we not the ones who program them and who decide when and where they should be used and what actions they may perform? In this view, computers are regarded as so-called 'standalone' machines that are under the direct influence of their inventors. It seems logical to think that the maker of a machine, or any other individual for that matter, should be able to pull the plug at any time.

Yet two reasons can be given as to why this view can no longer pass unchallenged. The first is that computers, like the chips developed by BrainGain and Warwick, are not isolated entities. They are in constant connection to large-scale networks which cannot easily be switched on and off—rather like the Internet. The latter is sustained by many different connections, commands and operations that are carried out simultaneously from many different locations, and it thus possesses a high degree of autonomy. 'Pulling the plug' in order to turn off the Internet is not a feasible option when these computerised systems are connected to huge networks that can no longer be traced back to a single user or operator.

A second reason for challenging traditional ideas about humans being in control of computers is the latter's increasing processing power. Anyone who regularly replaces their personal computer is aware of the fact that computer capacity increases at a fast rate. This increase is mostly to be found in processing power and memory size—that is to say, in computers' capacity to quickly store and access large amounts of data.

Nevertheless, the degree to which computers are able to understand and interpret spoken language as well as visual and auditory signals is still rather limited. The chess computer that beat Gary Kasparov, for instance, did not possess the grandmaster's knowledge or intelligence. In fact, all it did was run calculations on what would be its best next move by processing huge numbers of possible move sequences. This is not to detract from the largely untapped potential of computers to increase their ability to *learn*. By adopting evolutionary techniques which can be acquired at a much faster rate than is the case in natural evolution, computers may in time become worthy of being called intelligent creatures.

In a 2001 interview with German magazine *Focus*, the renowned British physicist Stephen Hawking, who is almost fully disabled as a result of a crippling muscle disease, warned that artificial intelligence can pose serious threats to humankind. According to Hawking, every 18 months the computational capacity of computers doubles, making it difficult for humans to keep up. There is a serious risk of computer intelligence increasing to the point where we can witness machines that are able to function with full autonomy. Hawking added the admonition that for this very reason, humans should try and match the development of computers. It is of the utmost importance for humans to try and maintain a biological superiority to electronic systems.

This point is also stressed by Kevin Warwick, who asks 'what is wrong with adding something that gives you extra capabilities?' As far as he is concerned, we can either let artificial devices determine our destiny for us or try and augment our own human intelligence. The creation of interfaces which link the human brain to computers will ensure that human cognition is supported by artificial intelligence, which should guarantee that humans remain masters of their artificial creations. However, as any hacker will tell you, once a network is in place it is difficult to stay in control of all the digital traffic. Therefore, Warwick's vision of the future is susceptible to concerns about the degree of control that can be exercised over networked brains. Warwick's answer to this concern is that, in a world of cyborgs, the notion of 'self' is likely to have a fundamentally different sense, thus rendering ideas of 'self-control' equally open to new conceptualisation.

Constructing Intelligent Computers

The development of intelligent machines first gained scientific interest in the mid-1950s with the advent of artificial intelligence (AI). Inspired by the development of computer technology, scientists commenced research on the construction of intelligent computers that were patterned after the human brain. Since little was known about the workings of the brain, it was hoped that the creation of intelligent computer systems would also provide us with greater insight into human intelligence. In the early years of artificial intelligence research, many believed that it would take 10–20 years before computers could be built that would be able to perform any task a regular human was capable of doing. This deadline was not met, and to this day a truly intelligent computer system has not been created—although a lot has been accomplished in the meantime.

One of the early focal points of AI research was the development of chess computers. It was believed that if computers could beat humans in chess, this would be an important step towards the development of systems that were truly worthy of the honorific 'intelligent'. The victory of IBM's computer Deep Blue over chess grandmaster Garry Kasparov in 1997 finally ended years of speculation about whether such an intelligent chess computer could really be produced. This accomplishment was considered a significant achievement in the history of AI, although it did not signify that computers now had the capacity to function independently and autonomously from humans.

Apart from the development of chess machines, research also focused on the construction of robots, particularly in Japan. Robots are first mentioned in the science fiction literature of the early 1920s as mechanical beings which resemble humans and may be used as replacements for human labour (in fact, the word 'robot' is derived from the Czech word for forced labour). Even as far back as the eighteenth century, various machines were constructed which mimicked human or animal behaviours—a famous example being an eighteenth-century Japanese doll that could offer tea to its guests. Japan, as will be seen in this chapter, is one of the countries at the forefront of the development of humanlike robots.

The first generation of robots consisted mainly of so-called 'service robots', which relieved humans of certain tasks that these robots could perform non-stop. A prime example are the robots used on factory assembly lines or conveyor belt systems, such as the type of robot used in the car industry. Service robots thus represent the first successful application of robot technologies. They can be deployed on tasks that humans experience as highly monotonous, that require high levels of precision or that could prove hazardous for living beings (such as the neutralisation of explosives).

A number of 'consumer robots' have also been created, functioning, for example, as robotic companions or playmates. Examples of this type of creation include Sony's robot dog Aibo (which is programmed to obey its master); Honda's Asimo (one of the first robots capable of walking upright, reminiscent of an astronaut in its looks, see Fig. 7.1); Paro (a therapeutic robot seal); the Tamagotchi (a digital companion which became popular in the early 1990s and has recently become available for smartphones); and finally the robotic housemate that is Robosapiens 2004.

These service and consumer robots exist to serve humankind; they are literally controlled and programmed by their human masters. In addition, an important part

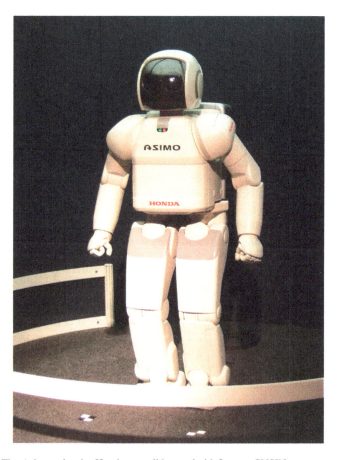

Fig. 7.1 The Asimo robot by Honda: a walking android. Image: GNSIN

of scientific research is geared towards the construction of self-improving robots that can function with some degree of autonomy. A central feature of this type of research is the interaction between humans and machines. Unlike service robots, these machines are programmed to acquire data by means of environmental interactions and to use these data to increase their range of capabilities. In this way a number of robots have been developed which possess eerily human characteristics. As we shall demonstrate, such robots evoke a sense of the eerie precisely because they further elide the distinction between robot and man.

In what follows, we will first introduce several examples of robots that imitate human facial expressions in order to communicate with their environment. Second, we discuss a number of robots that have physical attributes similar to humans, which they use to walk, jump, dance and so forth. Lastly, we cover a set of robots which are so much alike to humans in movement and appearance that they can almost pass for actual humans.

The Human Interface

The Kismet robot built by MIT in 1999 is best described as a kind of social robot with human features. Its main human quality is its propensity for facial manipulation (see Fig. 7.2). Kismet can interact with its environment through visual, auditory and proprioceptive sensing systems. This enables it to produce a wide range of facial expressions. For instance, Kismet can choose to convey looks of cheerfulness, calmness, sadness, disgust or anger. The way the robot behaves is modelled on very basic human reactions to external events.

The goal with which scientists have constructed Kismet is to find out whether a robot can be trained to adopt new behaviours through its experiences with the world. The behaviour adopted by Kismet (looking away when an object gets too close, conveying looks of glee, surprise or puzzlement) has not been predetermined, but is a consequence of its interactions with surrounding elements.

When no visual stimuli are presented to Kismet, its facial expression changes. It conveys looks of sadness and displays behaviour which we associate with loneliness, and consequently starts a quest for human companionship. By implementing various learning mechanisms, it is hoped that Kismet will continue to develop into a robot that possesses a degree of social intelligence and is capable of engaging humans in social interactions.

A comparable initiative is the iCat project by Philips, which aims to study man–machine interactions. The iCat robot possesses an internal camera which it uses to scan objects and facial expressions. Through a series of built-in microphones, iCat is able to detect sounds and pinpoint whence they originated, and thus to use those sounds as a platform for speech recognition tasks. It further contains sensors that register tactile input. Another key aspect of the robot is its ability to manipulate its facial expressions to make them strongly correspond to those of humans (see Fig. 7.3).

Fig. 7.2 MIT's Kismet robot can assume various facial expressions. It has visual, auditory and sensory sensors in order to interact with its environment. Image: Nadya Peek

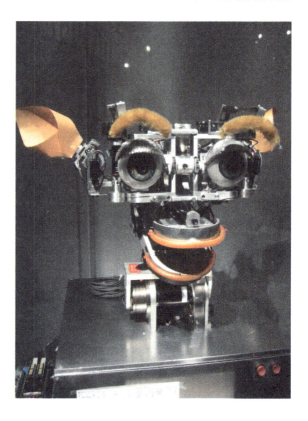

Whereas Kismet was a unique, one-off project, Philips aims to produce the iCat in larger numbers in order to facilitate future research. For this reason, copies of the iCat robot are being employed in several research locations. Surrounding the iCat is also an online community of users who exchange tips and experiences with one another. This creates an environment in which a collaborative development and improvement of its design is made possible.

Both the Kismet and the iCat project opt for a facial configuration which is clearly different from that of humans, yet which also maintains a correspondence of features. Both robots display emotions that are very recognisable to humans. Research has shown that the closer a robot resembles a human being in looks, the more critical humans become of its expression and behaviour. This is why both Kismet and iCat make use of the techniques of cartoon drawing. These robots, while clearly having non-human appearances, at the same time possess features which are easily recognisable as belonging to humans. Just as cartoon images, Kismet and the iCat robot come across with more conviction by displaying highly exaggerated expressions and emotions. Coming to our next example, we encounter a completely different type of robot, namely a robot which imitates humans not in expression but in movement.

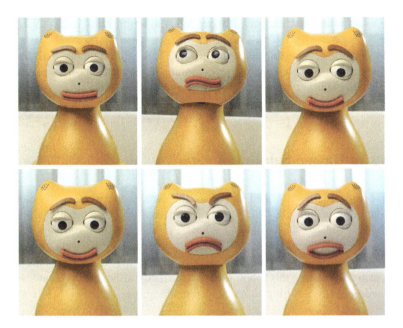

Fig. 7.3 iCat by Philips. Source: safeliving.wordpress.com/2007/11/20/telecare-session/icat-by-philips-expressions

Dancing and Jumping Robots

Along with communication robots that are geared towards interaction with humans, there exists a wide range of robots that move and dance, of which the Beatbot is the funniest to behold. It kind of looks like a yellow snowman or rubber duck which bobs along to the beat of different kinds of music. Children in particular get much excited by the sight of this robot, which does not just respond to music, but also interacts with its surroundings by responding to the movements of those in its vicinity.

A dancing robot is quite something to behold in itself. But to encounter robots that dance with each other and even flawlessly execute a complex Japanese choreography is truly a high point of modern robotics. Sony's QRIO robot is capable of such performances, as is documented in many videos that can be found on YouTube. Balancing itself on two legs, it can move both arms and legs in a graceful manner. Its appearance matches the prototypical look of a robot; it is not more than a steel construction with angular dimensions. Its movements, on the other hand, do reflect those of humans.

Dexter, a robot produced by the American company Anybots, is a closer approximation to the human musculoskeletal system. It is the very first robot to have the ability to walk on two feet, as well as being the first to be equipped with a flexible spine. Its walking motion is not exactly elegant; in fact, it is more of a shuffling trot than an actual walk. Dexter looks more impressive, however, when jumping, which it is capable of doing in a very realistic fashion. Because of its flexible spine, it is even able to get back on its feet without human assistance—a feat which few robots are capable of performing. However, for all its advantages, Dexter is also still very much a robotic being, as little to no effort was put into giving it a human appearance.

Robots as Human Substitutes

Japanese researcher Hiroshi Ishiguro raised the bar significantly when he produced an actual copy of himself. In 2006, after a successful trial with an android that looked like a well-known Japanese TV host, Professor Ishiguro developed a robot which was a realistic duplicate of himself (see Fig. 7.4). Viewing the moving images of Ishiguro's mechanical doppelgänger, one quickly understands all the enthusiasm surrounding this robot: it almost perfectly resembles its creator in both looks and behaviour. Sitting in its chair, breathing, gazing around while slightly

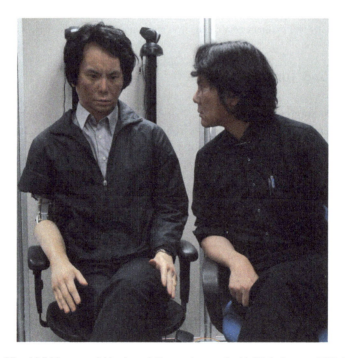

Fig. 7.4 Hiroshi Ishiguro and his doppelgänger, the geminoid HI-1. Image: ATR intelligent robotics and communication laboratories. Geminoid was developed by Hiroshi Ishiguro laboratory, advanced telecommunications research institute international (ATR). Trademark: Geminoid is a registered trademark of advanced telecommunications research institute international (ATR)

tapping its toes—the robot does everything we would expect from a 'real' human being. Every effort was made to create a robot that looked exactly like the original, and the result is a robot that comes across as highly convincing (as evidenced by its ability to put on a very impressive poker face).

The robot is operated remotely through Ishiguro's voice, body posture and facial movements, which are scanned and then reproduced by the robot with extreme precision. Professor Ishiguro can be miles away, and still the robot is capable of receiving signals and of giving a perfect imitation of Ishiguro's actions. Although this might come in handy, say, when being an hour's drive away from the location where one's presence is required for teaching, it does not mean, of course, that the robot can give lectures independently. Then again, that is not the point of Ishiguro's research, which sets out to study man–machine interactions. A more interesting question would be to what extent people feel themselves in the presence of Ishiguro when dealing with his android, or, conversely, to what extent they engage the android in communication in a manner that is natural when dealing with other humans. What makes us recognise other people as fellow human beings? Is it a question of looks or behaviour? In the end, these are some of the deciding factors for determining whether or not humans can be engineered.

Through his experiments, Ishiguro has discovered that humans who only get so much as a brief glimpse of a stationary robot will have no problems identifying it as such. However, when the robot is also making slight movements, the majority will believe they are dealing with a fellow human. Especially involuntary movements, such as tilting the head, rotating the neck while gazing around or making small hand movements, are identified as contributing elements to the persuasiveness of an android robot. Although the ability to speak is also significant, it is mostly small body movements which give the robot a lifelike impression. The so-called 'body movement factor' thus seems to be of greater importance than speech for achieving realistic communication between androids and humans.

Overall, the clips of Ishiguro's doppelgänger come across as highly convincing. In addition to Professor Ishiguro's clone, another android was created by scientists which was designed to look like Einstein, and there are several more specimens that walk, talk and move like humans. All the same, these androids are also reminders that a lot of work is yet to be done. Ishiguro's android, for example, cannot walk or move on its own but is fixed to a chair. It is also fully dependent on Ishiguro's presence to perform some of the more complex tasks of which it is capable. This goes to show that android robots are still far from being regarded as fellow human beings.

Enhanced Humans

On the one hand we have provided a sketch of the rise of cyborgs that have the ability to communicate via fixed computer networks, and on the other hand we noted the increase in android robots that show such a great resemblance to humans that we might even become convinced of their humanity. What does this mean for the way in which humans and machines will coexist in the future? Two other questions are relevant to this issue. (1) To what extent is it technologically *feasible* for humans to become cyborgs engaged in large-scale communication with computer systems, and for android entities to function on an equal basis next to humans? (2) To what extent is it socially *desirable* to construct such android robots and cyborgs? There remain wide differences of opinion on both these questions.

A large number of experts are convinced that within the next 50 years computers will advance to the point of existing on an equal footing with humans. They argue that computers will one day display the same behaviour as humans, communicate in the same way as we do and be capable of showing human emotion. According to these scientists, it is irrelevant whether these emotions are of the same nature as human emotion or represent a form of trained behaviour. The Ishiguro robot shows that technology has progressed significantly, but it is also demonstrative of the fact that we still have a long way to go. Computers are not yet capable of thinking independently or of understanding casual utterances of language, nor do they have the ability to participate in social activities. Hence, a real breakthrough on these terrains is not yet on the horizon.

Scientists like Warwick and Hawking are nonetheless convinced that someday advanced computer programmes will not only equal the abilities of humans, but will even be capable of doing much more than us. They warn that if we do not watch our step, it is a real possibility that robots will eventually take over.

This potential loss of human control over machines is the central theme of the 1999 movie *The Matrix*. Kevin Warwick views its narrative as being not just fictional, but as representing a possible vision of our future. The film is set in the year 2199, at which time machines have defeated mankind, yet humans are not themselves aware of this situation. They are trapped in a virtual reality, a computer-generated dream world that resembles the society of the late twentieth century. Human bodies are carefully preserved in large incubators in order to supply the system with energy. According to Warwick, the thought that humans will serve as an energy source for computers is evidently undesirable, but not entirely unthinkable.

Not everyone shares his belief in the technological possibilities for realising the wonderful yet often apocalyptic visions of science fiction writers. A large group of scientists are still quite sceptical in their opinion on the question of whether machines can take the place of humans. According to this view, which is put forward by, among others, cognitive scientists and philosophers, the chances of us ever attending lectures given by Ishiguro's replica are very slim.

One of the most influential philosophers in the field is the American philosopher Hubert Dreyfus. Dreyfus does not find it surprising that no computers exist which can match the human brain. Intelligent systems are built around the notion that the human brain functions rationally, and that it structurally processes and reproduces information in a manner comparable to the algorithms utilised by chess computers. Dreyfus believes this to be an incorrect assumption. Humans do not apply rules when deciding how to act in a given situation; most persons probably not even do so when playing chess. The brain operates on intuition and experience. This bears little connection to the mathematical processing of pieces of information, according to Dreyfus.

Support for his theory comes from Ab Dijksterhuis, professor of social psychology at the Radboud University Nijmegen in the Netherlands. Dijksterhuis created a furore in 2007 with his contention that humans 'think with feeling'. According to him, conscious reflection is much overrated. We think that our consciousness is called upon when making important decisions and that it defines our intelligence. Yet most choices—even wise ones—are made subconsciously. The way we think and act is partially determined by the circumstances—it is *situated*, as Philip Brey calls it. Our situatedness is context- as well as location-dependent. It cannot be preprogrammed and can thus never be simulated by a computer.

The second question of whether or not it is culturally desirable to improve ourselves touches on an issue which is discussed in almost every chapter of this book: namely, to what extent do we want to enhance humans? The answer to this question is closely linked to what meaning we give to the word enhancement. A common distinction is made between *healing* and *enhancing*. We normally do not object to the idea of healing, nor do we object to the development of therapies or expedients for helping patients with an illness or ailment of sorts. But as soon as the debate turns to questions of enhancement, it becomes more contentious. A fine line exists between enhancement and healing. Glasses, for instance, are viewed as a means to compensate for deficiencies of eyesight. Yet telescopes are regarded as instruments that belong to the category of enhancement (see Rose 2006). But are not both examples of human adaptation and even modification?

It is often stressed in debates on this topic that humankind has always been involved in attempts at self-improvement. According to the American philosopher and ethicist Arthur Caplan, those who favour the idea that human nature is invariable and that new technologies will turn it into something unnatural are guilty of committing a basic fallacy. Such reasoning is based on the idea of the existence of a natural state in which humans can, and by necessity must, exist. According to German philosopher Peter Sloterdijk, the question of whether humans ought to use technology for self-enhancement is no longer a valid one, as humans have already been strongly shaped by technology. Human evolution is entangled with the evolution of technology, as the dissertation of Pieter Lemmens, a Dutch philosopher, tells us. Humans are deeply reliant on technology, whether in the form of 'hard' appliances such as computers and machines or 'soft' techniques like education and other forms of pedagogy and socialisation.

If we are already striving to improve our intelligence, skills and health via a system of education—including health education—then why should we not also implement new technologies for this purpose? This is the question advanced by British philosopher John Harris. If parents who are well-off have the opportunity to send their children to private schools, then someday they might also have the opportunity to augment their children's brain capacity. Is there a difference?

As we noted earlier, Kevin Warwick also sees the potential upside to the technological enhancement of humans. He joins the philosophers we mentioned earlier in seeing little difference between this kind of enhancement and other forms of human improvement or healing.

> People with pacemakers and cochlear implants [an electronic implant inserted into the ear; used to restore a sense of sound for patients whose hearing is lost or damaged] are getting a benefit from technology. What is wrong with adding something that gives you extra capabilities? (*The Guardian* October 4, 2001)

John Harris dares to go a step further. He not only thinks that human enhancement should be allowed, but even regards it as our moral duty. Thus, natural selection, an arbitrary process to which humankind has been subject for centuries, could be replaced by a more rational and deliberate selection process:

> This new process of evolutionary change will replace *natural selection* with *deliberate selection*, *Darwinian Evolution* with *Enhanced Evolution*. (Harris 2007, p. 21)

Human Machines and Mechanical Humans?

What does this increasing interest in the development of cyborgs and android robots mean for future societies? What sort of novel relationships and new codes of conduct can or should we expect? In an attempt to provide answers to these questions, the American author Asimov articulated a series of rules or laws in his novels which he felt all robots should obey. He called them the three laws of robotics:

1. A robot may not injure a human being or, through inaction, allow a human being to come to harm.
2. A robot must obey orders given it by human beings except where such orders would conflict with the First Law.
3. A robot must protect its own existence as long as such protection does not conflict with the First or Second Law.

A fourth or 'zeroth' law was added later, which takes precedence over the other three:

4. A robot may not harm humanity, or, by inaction, allow humanity to come to harm.

These laws are clearly designed to stress the control that should be exercised over robots. Robots exist to benefit mankind, not the other way around. For Asimov, humans and machines represent two classes of beings that are clearly distinct from one another.

In the film *The Matrix* this distinction between man and machine is less clear-cut. Neo, the main character in the movie, is part of a group of humans who set out to destroy the Matrix's powerful machines by running their own competing computer programmes. Neo and his friends navigate between the virtual world of the Matrix and the 'real world' outside the system. Although at first glance *The Matrix* appears to feature a battle between humans and machines, Neo and his crew frequently hook up their own brains to the Matrix computer system in order to gain access to the virtual reality in which other humans dwell. This reveals the existence of a close interrelationship between humans and machines. The concept of individuality in the Matrix thus also has quite a different sense.

As we have highlighted throughout this chapter, it is common practice for today's scientific community to speak not of man *or* machine, but rather of man *and* machine. The line between the two categories will in all likelihood become increasingly blurry. Androids are becoming more and more lifelike, whereas humans, wired to large computer networks, show ever more traits that are typical of cyborgs. Scientists such as Warwick and Hawking not only see this development as defining our future, but they advocate an active role for humans as custodians of that future. The human brain ought to be enhanced by technology in order to retain control over new forms of artificial intelligence and artificial life.

The most important question regarding the future is not whether androids and cyborgs will be welcomed by society, but how society must try to redefine the relationship between the two categories. If the distinguishing characteristics of humans and machines continue to soften, Asimov's laws of robotics will no longer hold valid. The answer to the question of how android robots and human cyborgs may coexist—as illustrated by Koops's contribution on human rights (Chap. 12, this volume)—presupposes new social, legal and economic orders. The question is already relevant today. It is only a matter of time before it becomes urgent.

Bibliography

Books and Articles

Bostrom N (2006) Welcome to a world of exponential change. In: Miller P, Wilsdon J (eds) Better humans: the politics of human enhancement and life extension. Demos, London, pp 40–50

Brey P (1997) Hubert Dreyfus, mens versus computer. In: Achterhuis H et al (eds) Van stoommachine tot cyborg. Denken over techniek in de nieuwe wereld. Ambo, Amsterdam, pp 43–68

Caplan A (2006) Is it wrong to try to improve human nature? In: Miller P, Wilsdon J (eds) Better humans: the politics of human enhancement and life extension. Demos, London, pp 31–39

David L (2007) Intimate relationships with artificial partners, diss. Maastricht, Maastricht University Press

Dijksterhuis A (2007) Het slimme onbewuste. Denken met gevoel. Bert Bakker, Amsterdam

Harris J (2007) Enhancing evolution: the ethical case for making better people. Princeton University Press, Princeton

Lemmens P (2008) Gedreven door techniek. De menselijke conditie in de biotechnologische revolutie, diss. Nijmegen. Box Press, Oisterwijk

Rose N (2006) Brain Gain. In: Miller P, Wilsdon J (eds) Better humans: the politics of human enhancement and life extension. Demos, London, pp 69–78

Warwick K (2002) I, Cyborg. Century, London

Project Websites (In Order of Treatment; Accessed September 2012)

www.kevinwarwick.com/
www.wired.com/wired/archive/8.02/warwick.html
www.nici.ru.nl/braingain
paro.jp/english/index.html
www.ai.mit.edu/projects/humanoid-robotics-group/kismet/kismet.html
www.research.philips.com/technologies/robotics.html
www.hitech-projects.com/icat/index.php
www.cs.cmu.edu/∼marekm/projects/beatbots/
www.sonyaibo.net/aboutqrio.htm
www.anybots.com/
www.irc.atr.jp/∼ishiguro/
www.pinktentacle.com/2006/07/geminoid-videos/
www.kewego.nl/video/iLyROoaftIMW.html

Chapter 8
Human Enhancement, Evolution and Lifespan: Evolving Towards Immortality?

Simon Verhulst

Abstract Life expectancies of closely related species can differ substantially from one another. Such contrasts illustrate that life expectancy is not an inert quality of a biological system, but an evolved trait. A key factor in the evolution of lifespan is the extent to which individuals of a population are subjected to mortality factors that are largely beyond their control, such as the likelihood to die from disease, predators or starvation. A well-supported prediction in this context is that ageing rate will evolve to slow down when the risk of such extrinsic mortality decreases. For evolution to occur, only a very small number of criteria need to be fulfilled, and it is argued that humans still fulfil these criteria and that contemporary human populations thus continue to evolve through natural selection. The environmental change experienced by Western human populations is such that extrinsic mortality has been decreasing for many decades. This is due to changes in the external environment (cars, houses, industrialised agriculture), but also to changes in the body itself due to better nutrition and medical interventions. Based on the preceding points, the prediction is made that when the current low extrinsic mortality rate persists, human lifespan will evolve to become even longer than it is today.

Translated by Liseth Wielema.

S. Verhulst (✉)
Behavioural Biology, University of Groningen, Nijenborgh 7,
9747 AG, Groningen, The Netherlands
e-mail: verhulst@rug.nl

> If I'd known I was going to live this long, I would have taken better care of myself (Eubie Blake, an American musician who lived to be a 100 years and 5 days)

Turbulent technological and biomedical developments have given rise to speculations about human enhancement and the ways in which it will develop in the coming years. Many of these developments aim to repair damage to the body, striving to enhance bodily functions to an acceptable level or—in a somewhat broader interpretation of 'repair'—to the ideal level. The increasing ease with which the human body can be altered may seem like a break with history, in which humans have gradually developed as a product of evolution through natural selection. However, as will be explained in this chapter, this is not necessarily the case. The current tinkering with the human body deviates less from what humans have been doing for millions of years than appears at first sight.

In addition to developments in biomedical engineering, many aspects of the environment we live in are being 'engineered' as well. By controlling our environment to an important extent, we are less at the mercy of the elements. Thus, vast parts of the world are becoming increasingly safe and death from starvation or hypothermia increasingly rare. The increase in medical possibilities and environmental quality is not only beneficial in the short term, but also has long-term effects on human evolution. This chapter addresses the latter issue: what are the effects of human interventions in the body and the environment on human evolution, in particular on the development of lifespan? In this context, I will argue that the human life expectancy will increase through evolution, in addition to the already-observable effects on life expectancy of environmental factors such as improved nutrition and hygiene.

Human Life Expectancy in Perspective

At present, the maximum lifespan recorded for humans is 122 years, with a mean age of around 80 years in Western countries. How does this relate to other species? Frequently used research subjects such as the fruit fly and the worm *C. elegans* have a lifespan of mere weeks, which is short, even for flies and worms. Many invertebrates live substantially longer; some insect species can live for more than 20 years, and shellfish have been discovered of over 400 years old. This astonishing variation in lifespan can also be observed in vertebrates. On the one end of the spectrum, we find a small Australian fish species which does not live beyond 59 days, and on the other end we encounter tortoises of at least 175 years old and whales that are likely to live over 200 years—and the latter numbers are merely based on the few individuals whose age we happen to know. The ultimate survival champions, however, are trees. While hundreds-years-old oaks are impressive, they are not very special: they pale by comparison to the bristlecone pine, which can live up to 6,000 years!

8 Human Enhancement, Evolution and Lifespan

This brief overview illustrates that over the course of evolution a wide range of life expectancies has developed, in which humans do not hold a very notable position (yet). More remarkable is the fact that the life expectancies of closely related species can differ considerably. By keeping the queens of different species of termites in captivity for extended periods of time, we have determined their approximate maximum lifespans, which vary from several years up to 25 years. Examples such as this prove that maximum lifespan is not an inert quality inherent to a biological system, in the sense that a general genetic blueprint barely allows for variation. On the contrary, evolution's potential to change lifespan proves to be tremendous—offering interesting prospects for our species.

Evolution and Life Expectancy

What factors determine the maximum lifespan of a species, and why does this differ so much between species? A possible explanation can be found in species-related differences in an individual's risk of one day suffering a fatal incident. The term 'extrinsic mortality'—incidental death—is used to denote the risk of mortality caused by factors largely outside of an individual's control. In other words, behavioural and physiological changes in the natural repertoire of a species have no effect on its extrinsic mortality risk; it mostly boils down to 'bad luck'. Logically, an elephant has a smaller risk of mortality from external causes than a mouse, as the latter is more likely to fall prey to, for instance, a predator or a flood. Thus, variation in the extrinsic mortality risk partly explains the variation in mean lifespan between species.

However, the direct effect of variation in extrinsic mortality is only part of the story. A mouse's lifespan is limited to 2–3 years, even if its living conditions are optimal—permanent access to nutrition and an agreeable environmental temperature—but an elephant can live up to 50 years under the same circumstances. Apparently, mice and elephants also differ in 'intrinsic mortality': death from physiological failure. Extrinsic mortality ('bad luck') only explains part of the variation in mean life expectancy as observed in the wild and none of the variation in maximum life expectancy—which is entirely determined by intrinsic mortality.

One approach to account for the variation in maximum lifespan is to investigate ageing mechanisms that cause intrinsic mortality, such as cancer, cardiovascular diseases or other diseases related to ageing. By attempting to understand the underlying mechanisms, researchers from the field of biology and beyond confine themselves to the 'how' question with regard to variation in lifespan or other phenomena. Evolutionary biology adds a dimension by asking the 'why' question as well as the 'how' question: why have certain genetic variants prevailed in the course of evolution, while others are now extinct? To illustrate, the question of why polar bears have white fur can be approached in different manners. The answer to the 'how' question focuses on the genetic and physiological mechanisms that determine fur colour. The answer to the 'why' question, on the other hand, is fundamentally different, as it addresses the consequences of white fur colour on evolutionary success—'Darwinian fitness'—in relation to other possible fur colours—for example with respect to thermoregulation. In other words, they are trying to understand, from an evolutionary perspective and on the basis of differences in evolutionary success, why polar bears are white—and not brown or black like other, closely related bear species. Variation in lifespan is particularly interesting from the 'why' perspective, as death almost always decreases an individual's Darwinian fitness. If a random tortoise taken from the Galapagos Islands by Charles Darwin can live to be 175 years old, why cannot we?

Numerous theories have been put forward to explain how ageing arises in individuals, but there are only few evolutionary theories that explain why ageing has evolved at all and how it differs between species. One important theory, the mutation accumulation theory, postulates a process that in all probability does occur, but it remains inconclusive whether this process is quantitatively relevant. Therefore, this chapter will focus solely on the trade-off theory, which is supported by relatively much evidence.

The trade-off theory of ageing is based on the idea that the use of time, energy and nutrients is limited. Since, from the perspective of evolution, life exclusively revolves around procreation, there are only two ways to use raw materials advantageously: for procreation in the present (children and grandchildren) or for survival for the purpose of procreation in the future. There are many ways to invest in survival, such as investing in DNA repair, for which over 100 different mechanisms are known, and building up a fat reserve in order to survive the winter. Due to limited availability, the use of resources for reproduction comes at the expense of resources available for the stimulation of survival. The trade-off theory assumes that the ageing process accelerates as fewer resources are used for sustenance, thus creating a trade-off between reproduction and remaining life expectancy.

This idea has been tested extensively by manipulating the investment in reproduction. I have personally tested this for free-living birds by reducing the number of young per nest, allowing the parents to invest less energy in raising their brood. This proved to have positive effects on the parents' survival and future procreation. Research of this nature shows that an increase in the number of young produced in a given season is at the detriment of the number of young in later seasons. Evolutionarily speaking, this is a form of accelerated ageing. Thus, the rate of ageing increases as more is invested in reproduction.

In the context of the trade-off theory, the question of what causes the variation in ageing rate between species can now be reformulated: what determines the optimal distribution of resources between reproduction and survival? In essence, the answer to this question has already been found: the optimal distribution, and consequently the rate of ageing, is determined by the extrinsic mortality risk. The higher the risk of death by external causes, the more beneficial it is to invest in reproduction rather than survival, and the higher the rate of ageing. This elegant result can be easily explained by drawing an analogy with investing in a pension. If there is only a small chance of reaching retirement age, there is little incentive to invest in a pension, and money is best spent in the present. Conversely, the chance of living beyond the retirement age by a considerable number of years provides an incentive to invest substantially in a pension, so as to save money for future enjoyment.

This insight gives rise to the interesting prediction that there is a link between extrinsic and intrinsic mortality: species that have a low life expectancy due to external causes will, on average, age faster and live relatively short lives even if they are housed under optimal conditions. This is perfectly in accordance with the comparison between the mouse and the elephant mentioned above.

Evolution and Humans

Darwin's most important contribution to biology is the concept of evolution through natural selection. This concept excels in its simplicity, attesting that evolution is inevitable if only a small number of conditions are fulfilled. What is evolution, under which conditions does it occur, and do humans still meet these criteria in the current era of human enhancement? If the answer to the last questions is 'yes', then humans are still evolving. Or has our living environment become so artificial that biology and evolution no longer play a part in our lives?

Evolution is at present defined as the change in genetic composition of a population (*at present* because genetics as we know it did not yet exist in Darwin's time). Appearance, physiology and behaviour—the phenotype—are the result of an interaction between the genome and the environment. Phenotypic variation within a population can partly be explained by environmental variation and partly by genetic variation. The relative importance of genes and environment is dependent on the trait in question. New genetic variation is the result of errors (mutations) in the copying of DNA for the production of egg cells and sperm cells. There are many different types of errors, such as the duplication or deletion of whole segments or a mutation in a single base pair—the smallest unit of DNA. Since these kinds of errors can never be entirely avoided, mutations occur in all organisms, including humans.

Mutations are random—that is to say, unregulated. For this reason, most mutations result in a small decrease in the quality of the carrier, in the sense that the carrier will produce less offspring than would have been the case without the mutation. This negative effect is not surprising, in the same way that it is hardly surprising that changing a random note in an existing piece of music is likely to render it slightly less harmonious. Testimony to this principle is that mutations in humans are usually detected because they decrease our health in one way or another. Naturally, mutations can also have a positive effect, in which case the carrier will produce slightly more evolutionarily successful offspring than non-carriers—just like a random change can sometimes improve a piece of music. As genetic variants influence the number of produced offspring, the genetic composition of the next generation will automatically be different—the species is evolving!

Since this variation in number of offspring applies to humans as well, and since new genetic variants continually emerge, it is inevitable that human beings are also selected on the basis of positive genetic variants. If selection is very strong, in the sense that there are large disparities in success, natural selection at a genetic level is easily detectable. For instance, prostitutes in Nairobi (Kenya) are constantly exposed to HIV through unprotected sex, and most of these women will eventually contract AIDS. Some women, however, appear to be immune, as they do not develop AIDS despite years of continuous, frequent exposure. If this resistance has a genetic basis—for which there is some evidence—there will be a strong selection pressure favouring this genetic variant, which will accordingly spread quickly throughout the population.

Evolution and Environment

What type of genetic variant is favoured by natural selection depends on the environment, as this determines the optimal features of an organism. A case in point is resistance to HIV infection, as discussed in the previous section. A genetic variant with this effect may be selectively neutral in a world without HIV or may even affect the Darwinian fitness negatively, but it has a strong selective benefit in areas where HIV is widespread. Understanding and predicting the evolution of an organism is therefore only possible if its environment is taken into consideration.

A remarkable feature of the human species in comparison to other animal species is the extent to which we modify the environment we live in. To a more moderate extent, this type of behaviour also occurs in other animal species and is known in ecology as 'niche construction'. An impressive example of this phenomenon are the enormous nests, providing safety and a constant climate, in which termites spend a considerable part of their lives. An exciting consequence of niche construction is that a species influences its own evolution—after all, evolution is dependent on the environment. This way, human beings probably exert a relatively strong influence on their evolution. Consider the following example. Comparative research between tropical and non-tropical animal species has shown that tropical species are less capable of generating extra heat when placed in a cold environment. The human ability to protect ourselves from cold by means of fire, clothing and housing may, in a similar way, have resulted in a decrease in our ability to generate body heat in the course of evolution. Accordingly, it can be argued that through the process of evolution we have become (or have remained) a tropical species, since we have adjusted our environment in such a way as to be no longer dependent on our ability to generate body heat.

The Body as Environment

From the perspective of the genome, there is only a modest difference between the modification of the environment we live in on the one hand and the 'enhancement' of humans by means of biomedical interference, improved nutrition and protection against diseases on the other. From this point of view, the body is part of the genome's environment.

Indeed, from the perspective of a gene the rest of the genome is also part of the environment, in the sense that the modification of a single gene is nothing but a change in the environment to all other genes, and therefore not fundamentally different from 'real' environmental changes. To a gene, the body is merely a disposable product designed to multiply the gene—in the same way that a chicken can be regarded as a means used by an egg to create a new egg.

In the context of evolution, the recent biomedical developments, including genetic manipulation, and the environmental manipulation that has been taking place for centuries are closely related. As far as the gene is concerned, both are cases of environmental manipulation. Therefore, when we have a lesser need to generate heat, it makes no difference to the genome whether this has been achieved by means of a jacket or a coat of body hair. Genes have no morals; the only thing that matters is the effect on procreation, regardless of how that effect is achieved.

Human Enhancement: Through Evolution?

Human beings live in a self-created environment which, in most parts of the world, has changed substantially during the past centuries, creating a large difference between the environment we currently live in and the environment in which our species has spent the majority of its evolutionary history—a difference that is rapidly increasing due to technical and biomedical developments. In the meantime, human evolution continues as humans are continually adapting to their ever-changing environment. Since the mechanism behind evolutionary processes is universal to a considerable extent, it is possible to make well-founded predictions regarding the human evolutionary response to key aspects of our altered environment.

An important consequence of this altered environment and recent biomedical developments is that human life expectancy in the Western world has increased significantly. The mean life expectancy of Japanese women, for instance, is increasing linearly with 3 months per annum—thus, for every year a Japanese woman lives, her remaining life expectancy decreases with only 9 months! This increase in mean life expectancy is partly the result of a decrease in infant mortality, but also when only the life expectancy of elderly people is taken into account, a significant rise can be detected. This is illustrated in Fig. 8.1, which shows that the life expectancy of 70-year-old Swedish men has been increasing for over 150 years. For the last 50 years, the figure shows a linear increase of 1.8 months per annum, which gives reason to expect that this trend will continue for several decades to come.

What effect does our spectacularly increasing lifespan have on human evolution? This longer lifespan is partly the result of a decrease in mortality through factors beyond an organism's direct control. Paradoxically, this also includes many aspects of the environment in which an organism grows up and lives—these

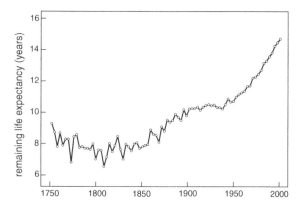

Fig. 8.1 Life expectancy of 70-year-old Swedish men. From: Finch et al. 2005

factors are within an individual's control, yet this control is negligible in comparison to the vast differences caused by the time and place in which an organism lives. As explained earlier in this chapter by means of the mouse–elephant comparison, life in an environment with lower extrinsic mortality induces an evolutionary shift towards slower ageing. It is worth saving yourself for later, as there is a good chance you will live to reap the benefits. Or, reversely, there is no reason to immediately 'burn yourself out'—after all, there is plenty of time.

The evolutionary extension of lifespan will occur in addition to (and because of!) the spectacular increase in lifespan that is already observable today due to human interventions in our living conditions. Exactly when and to what extent this evolutionary effect will take place is difficult to predict, but if extrinsic mortality remains low, it is certain that evolution will cause the average Western human lifespan to increase.

Side Effects

An extension of the human lifespan will have all sorts of biological and societal side effects, some of which can be predicted. A slower ageing process and longer lifespan are usually accompanied by a longer development time, denoting the age at which someone is able to have their first child. With regard to animals and plants, a longer development time usually means that by the time they start reproducing, organisms are bigger, stronger, more experienced or in possession of a more suitable territory. Thus, a longer development time has a positive effect on reproductive success—assuming an organism lives long enough to reap the benefits. It is likely that this effect occurs in humans as well, or at least during part of their lives. Having children at an early age hinders investment in education, and humans tend to become wealthier with increasing age. From this perspective, there is an advantage in postponing the start of a family, and this advantage becomes more beneficial with increasing life expectancy.

A disadvantage of extending the development time is that fewer years remain to be spent on reproduction, especially for women. From an evolutionary perspective, therefore, there is an optimal development time—that is to say, the ideal compromise between the pros and cons of extending development time. As lifespan increases, so does the optimal development time, since the price paid for postponing reproduction gets lower. Consequently, it is expected that an increasing lifespan leads to parents having their first child at a later age. This prediction is supported by the fact that there is a genetic link between the age at which women have their first child and their lifespan. In other words, there are genes that both determine that women will have their first child at a late age and, at the same time, that they will live a long life, making it evolutionarily easier to change both traits simultaneously.

An additional consequence of a longer lifespan is a further increase in the subjective value of human life. After all, a fatal car accident at the age of 50 has a different impact on a population in which the mean age of death is 55 than on a population with a mean age of 140. It seems likely that this will lead to extra risk minimisation. Perhaps children will all be taken to school in SUVs in the future, or will cars in general be considered too dangerous a means of transport?

An extended lifespan also has financial consequences. Given a certain level of productivity and consumption, an increase in lifespan means that either humans have to work more during their lives—for instance by retiring at a later age—or pensions will inevitably decrease. This issue is already being addressed; as of recent, pension funds have started taking into account estimates of the increasing life expectancy, and as a consequence have had to increase their reserves to be able to meet their long-term obligations.

Evolving Towards Immortality?

Making predictions with regard to human evolution appears to be somewhat of a gratuitous exercise—after all, the predictor will not live to see the time when the prediction will or will not come true. Fortunately, it is possible to further substantiate the prediction that evolution will add extra years to our lifespan. Assuming that the recent increase in lifespan will not be reversed (in spite of economic crises, multi-resistant bacteria and other plagues), only a few conditions have to be met in order to expect an evolutionary lifespan extension. In short, these conditions require (1) there to be genetic variation in lifespan and (2) for a longer life to increase the number of evolutionarily successful offspring.

With regard to genetics, it is particularly important whether a part of the variation in life expectancy is caused by genetic variation, or, in other words, whether human beings who live a relatively long life do so, to a degree, due to the composition of their genome, and not exclusively on the basis of environmental factors. After all, only when genetic variants exist that allow for a longer life can evolution lead to a longer lifespan. The influence of genes can be broadly determined by investigating the degree of correlation between the lifespans of relatives (see Chap. 5, this volume). As genes become more important, the lifespans of relatives will be more similar due to the genes they have in common. Calculations based on similarities in lifespan among relatives indicate that at least a part (15–30 %) of the variation in lifespan has a genetic cause. Moreover, the significance of genetic variation often depends on the environment, and does so in such a way that genes are more influential when the environment is of high quality.

In addition, genetic variation also appears to be more significant when premature death, such as death before the age of 50, is not taken into consideration in the analysis. Since a cohort is only useful for analysis when the majority of its members is deceased, genetic analyses so far have been conducted with data of individuals who were born a long time ago, for instance before 1900. It seems probable, therefore, that in the future genetic variation will prove to be more influential now than it was a century ago, as the quality of the environment we live in has increased over the past decades. In any case, it is already evident at present that genetic factors are important enough to enable evolution of lifespan.

Individuals who carry genetic variants that cause them to live longer will only enjoy an evolutionary benefit if they use this extra lifetime to increase their reproductive success. It is not immediately clear whether a longer lifespan also augments the number of successful offspring—especially in women, who do not bear children after menopause. Nevertheless, there are reasons to expect that, even now, persons with a relatively longer lifespan can produce more successful offspring. First, lifespan extension creates the opportunity to be used for procreation. New and existing genetic variants capitalising on this opportunity will be at an advantage and will therefore spread in the population—as will happen sooner or later.

Second, there is an evolutionary advantage to living longer, also for women who have already passed menopause. This is linked to the fact that the number of children in itself is not the most important factor, but that evolution selects on the total number of descendants in a more distant future. It is more valuable to have 5 children who produce a total number of 15 grandchildren than it is to have 10 children who produce a total number of 10 grandchildren. Thus, conceiving or bearing children is not the only way to produce more offspring. An alternative is to help your children in raising their children (your grandchildren), who are also carriers of your genes; it has been repeatedly shown that children's reproductive success is increased by the presence of a grandmother. A longer lifespan can therefore be evolutionarily advantageous even without personally producing extra children.

Conclusion

The environment we live in is governed by humans to such an extent that it seems justified to question the degree to which evolution through natural selection can still play a role in the development of human beings. The answer to this question is equivocal. On the one hand, it seems probable that selection is more tolerant to some parts of the genome now than in the past. For instance, the selection on the genes that code for our eye lenses is likely to have become more lenient since the invention of corrective glasses. On the other hand, the evolutionary pressure favouring a longer life, such as the 'grandmother effect', has increased natural selection on genes that determine lifespan. Even in the comfortable world we now inhabit, evolution runs its course. In fact, due to the recent, rapid changes in our environment, it is likely that we are currently evolving faster than ever.

Turbulent technological and biomedical developments along with improved nutrition and living environment enable us to live longer. This development is an ongoing process, and interventions in individual humans are getting more and more invasive, currently culminating in the possibility of genetic modification. This may seem a fundamental break with our past, yet from an evolutionary perspective it is a natural progression from previous developments. After all, when the body and the rest of the genome are merely considered to be parts of the environment—as is realistic from a gene's perspective—then there is nothing new under the sun. To a gene, there is no substantial difference between the invention of the wheel, putting on sneakers or genetic manipulation to prevent flat feet; from an unmanipulated gene's point of view, these are all environmental factors. Even in the biomedical age, evolution will run its course.

A surprising side effect of the extended lifespan that comes with the biomedical age is that evolution is likely to add to this effect. Consequently, our descendants will live even longer than we would expect solely on the basis of technological developments. Humans can therefore be enhanced in an unexpected way, in the sense that humankind will keep evolving as a result of human interference in its environment—including the body. At the same time, this is an entirely uncontrolled process. That is to say, while an increase in lifespan by means of evolution is predictable, we cannot influence this process, neither with regard to the process itself, nor with regard to the rate at which it occurs.

Immortality, however, will not be achieved, if only because the chance of a fatal accident will never be completely eliminated. If fatal accidents were the only cause of mortality, we could live to be over a 1,000 years old. While that provisional maximum may seem unrealistic at present, we will slowly move towards attaining it. Perhaps, the first person to live 200 years has already been born?

Bibliography

Austad SN (1997) Why we age: what science is discovering about the body's journey through life. Wiley, New York

Dawkins R (1976/1989) The selfish gene. Oxford University Press, Oxford

Finch CE, Crimmins EM (2005) Response to comment on 'Inflammatory exposure and historical changes in human life-spans'. Science 308:1743b

Chapter 9
Opting for Prevention: Human Enhancement and Genetic Testing

Annemiek Nelis, Symone Detmar and Elske van den Akker

Abstract Fictional portrayals of our possible future, such as the Hollywood film *Gattaca*, often conceive of a world where the genetic profile of each individual determines opportunity. Parents select the best sets of genes for their children to make sure they will be as successful, smart and healthy as possible. To what extent is such a scenario credible? Can the concept of genetic prevention be used to control, and if necessary adjust, life, preventing a variety of diseases and the sufferings these diseases can cause? And how realistic is it to offer techniques such as genetic screening for hereditary diseases to the population at large? These decisions have to be taken based not only on ethical, but also on economical considerations.

At the beginning of the movie *Gattaca*, the main character—Vincent—describes how he was born from an act of love between his parents. Vincent is born in a time when the selection of qualities and capacities that are genetically passed on from

---

Translated by Jenny Shelepov.

A. Nelis (✉)
Dutch Safety Board, P.O. Box 95404, 2509 CK, The Hague, The Netherlands
e-mail: a.nelis@safetyboard.nl
www.safetyboard.nl

S. Detmar
TNO, P.O. Box 2215, 2301 CE, Leiden, The Netherlands
e-mail: symone.detmar@tno.nl

E. van den Akker
Department of Medical Decision Making, Leiden University Medical Center,
P.O. Box 9600, 2300 RC, Leiden, The Netherlands
e-mail: vandenakker@lumc.nl
http://www.lumc.nl/vandenakker

B.-J. Koops et al. (eds.), *Engineering the Human*, DOI: 10.1007/978-3-642-35096-2_9,
© Springer-Verlag Berlin Heidelberg 2013

parents to children is not left to chance. During his birth the nurses openly discuss the choice Vincent's parents made: 'Honey, you have made one mistake... The child won't be thankful'.

Immediately after birth, a small amount of blood is taken from Vincent's umbilical cord for his DNA to be analysed. Page after page of results are printed by the computer, and several times the machine starts emitting loud warning sounds. The nurses exchange meaningful glances, causing Vincent's father to ask whether there is something wrong. The voice-over—Vincent's voice as an adult—points out:

> Of course, there was something wrong with me. Not so long ago I would have been considered a perfectly healthy, normal baby. Ten fingers, ten toes. That was all that used to matter. But now my immediate well-being was not the sole concern.

From the genetic data it is evident that in the course of his life, Vincent has a 60 % chance of getting a nervous disease, a 42 % chance of suffering from a manic-depressive disorder, a 66 % chance of becoming obese, an 89 % chance of having ADHD and, worst of all, a 99 % chance of dying of a heart attack at an early age. His life expectancy is 33 years.

For the time being, Vincent's parents decide not to have another child, at least not until they can afford to select the best set of genes through genetic engineering. They do not want fate to decide again how well their offspring will do in society. Several years later, with the help of medical technology, Vincent's brother Anton is born.

Gattaca portrays a world in which an individual's genetic profile is crucial. Society revolves around genes and DNA, which means that wealth, religion and race are no longer the source of social inequalities; now, genomes determine to what social class an individual will belong. The movie shows how much pressure this system puts on the genetically fortunate, as well as on individuals such as Vincent, who have to manage with a 'lesser' genome. Human beings are determined by the selection of the right set of genes.

Popular movies such as *Gattaca* provide us with an opportunity to reflect on as-yet unknown futures. This can be done, for instance, by identifying and articulating the social consequences and normative questions that follow from the future use of genetic knowledge and technology. The focus is often on the desirability of such future scenarios: is this the kind of society we would want to live in? This chapter, however, takes a different approach. We will use the movie as a starting point to reflect on the question of who or what decides on enhancement technologies, and, subsequently, how probable it is that in the future we will find ourselves in a '*Gattaca*' kind of world.

It is typical of many science fiction stories, including *Gattaca*, to rarely give any thought to *how* technology spreads throughout society. Often, the opening scene depicts a world where smart technology is omnipresent; in other words, the technology is already an accomplished fact. In so far as the question of 'who or what decides on enhancement technologies' is an issue in science fiction movies and stories, they are nearly always all-powerful government organisations reminiscent of the novel *Brave New World* or individuals who try to conquer the world and rarely succeed, as in *The Boys from Brazil* (see Chap. 10, this volume). Thus, in most science fiction stories new technologies are a given, and where they came from remains undiscussed.

This chapter will demonstrate that the answer to the question of who decides where, when and how new and innovative technologies should be used is not straightforward. The development and use of new genetic technologies are determined by a series of choices, which are made at different locations, by various people and with the help of all sorts of methods.

In the following, a health economist (Elske van den Akker), a medical anthropologist (Symone Detmar) and a science and technology studies scholar (Annemiek Nelis) will successively share their knowledge from their respective fields of study. They will show in what ways choices surrounding genetic technology play a role in the societal introduction of technology and the articulation of social and ethical questions. All three are concerned with healthcare science, and in particular genetics. Their stories about choices are situated at various different locations and describe various different actors, values and social effects.

The Case: Enhancement by Means of Prevention

In the literature a distinction is often made between 'improving' human beings (enhancement) on the one hand and 'preserving or restoring' their health on the other (medical intervention). Improving is aimed at the enhancement of performance, such as an increase in strength, intelligence or stamina. Restoring is aimed at a return to health, which can be effected by means of medication, medical intervention or lifestyle adjustments such as exercise and a healthy diet. In addition, it is possible to use medical devices such as artificial lenses, artificial hips, artificial hearts, artificial pacemakers and shunts. In general, these devices focus on restoring health, but sometimes they even lead to improvement. The distinction between the two is not always clear-cut. Artificial lenses, for example, are of such good quality these days that after implantation an individual might have better sight than ever before. The artificial lower legs used by South African sprint runner Oscar Pistorius, who participated in the 2012 Olympic Games, have also been under discussion, as it was feared that they might provide an advantage over 'normal' legs.

In this chapter we will focus on a topic that cannot immediately be linked to either a form of recovery or a form of enhancement: the *prevention* of disease, pain and suffering. What does the idea of human enhancement mean in the case of prevention? Prevention assumes that it is possible to enhance human life in the sense that it can be controlled and influenced. Implicit in this aim to control is a normative starting point or ideal: the wish to have a society of mainly healthy or disease-free individuals. In other words, the prevention of diseases, pain and suffering is based on the same ideal as the enhancement of human traits. Both try to influence the twists and turns of fate that naturally govern human life.

The prevention of disease can be accomplished in two ways: by taking necessary precautions so as not to get ill—for example by means of medicines, a healthy diet and exercise—and by preventing the disease and its determining genes from being passed on to future generations. Thus, letting individuals with a disease or condition 'not be born' is another form of prevention. From the beginning of the 1970s, an increasing number of possibilities have gradually come into existence to diagnose severe and incurable diseases at an early stage—that is, before birth.

Genetic diagnostics provides an opportunity to terminate pregnancy at an early stage in the case of an incurable congenital or hereditary disorder. Preventive genetic diagnostics enables parents to make well-informed choices about their offspring. It does not, however—according to those who offer this technology— aim to limit the number of children who are born with an incurable congenital or hereditary disorder. Nevertheless, this is one of the possible consequences or effects of this preventive technology.

Preventive genetic technology can roughly be divided into two categories: technologies used for *screening* and those used for *diagnostics*. Both aim to establish the presence of a gene mutation. Genetic screening is offered by the government and is directed at target groups that generally do not have any complaints or symptoms. The Dutch Health Council describes genetic screening as the 'examination of individuals to systematically detect at an early stage whether they have a hereditary disease or a predisposition to develop one, or carry a predisposition that may cause a hereditary disease in their offspring, irrespective of the type of examination used'. The purpose of screening is to detect potential risks.

In the case of genetic diagnostics the initiative lies with the individual. On the basis of complaints and symptoms or the family anamnesis it can be determined whether a person has a genetic defect. Individuals or families generally contact a medical practitioner themselves, such as their GP or a clinical geneticist. Their aim is to prove the presence of a genetic mutation. The possibilities for genetic screening as well as for genetic diagnostics have increased significantly over the past few years.

Preventive genetic diagnostics comprises both prenatal and postnatal tests, which are performed before and immediately after birth, respectively. Prenatal diagnostics has many manifestations, including obstetric sonography and more invasive methods such as the amniotic fluid test (during the 16th week of pregnancy) or the chorionic villus sampling test (during the 12th week of pregnancy). The much-discussed technique of pre-implantation genetic diagnostics (PGD, see also Chap. 5) is used by parents who are known to be at increased risk of a genetic disorder. PGD is gene diagnostics combined with IVF. During this procedure, several egg cells are fertilised outside the uterus. Once these fertilised egg cells have divided into eight cells it is possible, with only a very slight chance of damaging the embryo, to remove one of the cells in order to establish on the basis of the embryo's DNA whether the genetic disorder in question has been passed on.

By means of PGD and prenatal diagnostics it is possible to determine before birth whether an embryo carries a severe and incurable disorder. In the case of prenatal diagnostics, when a disorder such as Down's syndrome or *spina bifida* is found, parents are given the choice to decide whether they want to terminate the pregnancy. In the case of PGD, only the healthy embryos are transferred back into the uterus.

According to the information pamphlets and handbooks for genetic counselling, the main purpose of screening before birth is to provide parents with a choice. In the first place, this choice concerns whether or not parents want to be informed of any hereditary or congenital disorders in their unborn child. If the foetus turns out to be affected, the parents have another choice to make: to terminate the pregnancy or carry it to term.

There are several options for postnatal screening, including the Guthrie test. The primary purpose of this test is to decrease the harm or the disease burden by starting treatment early. Moreover, if a child is diagnosed with a hereditary disorder, parents can take this into consideration for any next pregnancies. Parents are presented with the choice of whether they want their child to be screened for a number of disorders, as well as the question of whether they want to receive information on their carrier status. In this chapter, we will focus on different aspects of both preventive genetic diagnostics and preventive genetic screening.

The Economic Decision: Setting Priorities

Prior to the provision of screening, another choice has often been made: the social or political decision to apply these tests in clinical reality. Although new testing possibilities are mainly developed in clinical practice and are initially only applied on a small scale and on a limited group of parents, eventually it is policy makers, politicians and insurance companies—joined by ethicists, lawyers and health economists—who decide which tests may or may not be offered on a large scale. It is often up to politicians and policy makers to decide which preventive techniques will be offered or remain available to the public.

This can sometimes lead to fierce debates. A case in point is the discussion that broke out in the Netherlands regarding the expansion of embryo selection in PGD, which centred on the question whether parents with an increased risk of a genetic form of breast or colon cancer should be eligible for PGD (see also Chap. 5, this volume). While the Dutch Health Council had already issued a positive opinion on this matter and everything had been set up to start the procedure, there was disagreement in the political arena about the desirability of this development.

There are more technological possibilities for preventive testing than those currently applied to the public; not everything that is technologically possible is also implemented. Since we generally need to set priorities, choices have to be made—including financial ones. Healthcare is a scarce commodity, and so are preventive measures. If all technological possibilities that are developed would also be applied, healthcare would become unaffordable. Against this background, how are choices regarding new technologies made, and what kinds of considerations play a part in this process?

Within the economy of healthcare, a number of criteria are used in order to enable a transparent decision-making process with regard to the prioritisation of often innovative technological possibilities. These criteria are solidarity, necessity and cost-effectiveness. Solidarity refers to the principle that everyone is entitled to good health and good healthcare. In addition, medical interventions with a higher necessity—that is, interventions that satisfy high healthcare needs—are more eligible for reimbursement from the collective healthcare budget than are medical interventions that are considered less necessary. Finally, medical interventions that produce better outcomes against similar or lower costs are also more eligible for finance. This 'value for money' can be determined by means of a so-called cost-effectiveness analysis.

In cost-effectiveness analyses the costs of a medical intervention are compared to the costs of other interventions or the costs of already existing healthcare. The difference between these costs is then linked to the additional health gain the medical intervention provides. This gain is preferably expressed in terms of 'quality-adjusted life years' (QALYs). A QALY is defined as one year of good health, physically as well as psychologically. If a medical intervention causes an individual's lifespan to be prolonged with one year in good health, then the health gain equals one QALY. If the intervention does not prolong lifespan but does improve the quality of life from, for instance, two-thirds quality to full quality for a period of three years, the gain is also one QALY. For example, the costs of lung transplantation for one additional QALY gained amount to €82,462, whilst the use of Viagra 'only' costs €4,163 per QALY. The use of Viagra is an example of an intervention that will not prolong life expectancy, but will lead to an increase in quality of life. In the case of lung transplantation the effect is mainly determined by the extra years of life enabled by the transplantation.

Based on the information provided by the cost-effectiveness studies, it is possible to draw up a table of the different interventions ranked in order of cost-effectiveness, a so-called 'league table'. The league table starts with the medical intervention with the lowest costs per QALY and ends with the intervention with the highest costs per QALY (see Table 9.1).

Table 9.1 Example of a league table. Source: Rutten-van Mölken et al. (2000)

Treatment	Alternative	Extra costs per QALY gained (in Euros)
Viagra	Andoscat	4,162
Breast cancer screening	No screening	4,204
Liver transplantation	Conservative treatment	36,402
Heart transplantation	Conservative treatment	38,206
Lung transplantation	Conservative treatment	82,462

Euros based on the price level in 1995

Based on the assumption that policy makers aim to maximise health gain within the limits of the available healthcare budget, a league table can provide the necessary information to select the optimal combination of medical interventions. Theoretically, the medical intervention with the lowest costs per QALY should be given priority by the government or healthcare providers. Next, it should be established whether the budget allows for the introduction of another medical intervention, and if this is the case, second priority should be given to the next intervention on the league table. This would then be repeated until the available healthcare budget is spent.

While this sounds good in theory, and although it seems natural that the selection of new innovative technologies should be based on a rational profit model, in practice the outcome of the cost-effectiveness analysis only plays a limited role. Decisions on the introduction of medical interventions are often not taken on the basis of strict criteria, but are also the result of other considerations and processes, in which factors such as the media, the industry and the lobbying of patient groups play a prominent part. In light of the economic principle of rationalisation, several initiatives have been put forward to argue for a greater role of cost-effectiveness analyses.

In England and Wales, cost-effectiveness analyses have been used more frequently since the establishment of the National Institute for Clinical Evidence (NICE) in 1999. The reasons for the foundation of NICE were the explosion of knowledge and technology in the previous decades and the related rise in costs of the National Health Service (NHS). One of NICE's tasks is to assess the effectiveness and cost-effectiveness of hospital care. Partly on the basis of these results, it is then decided whether or not to retain or introduce a medical intervention.

Moreover, in order to ensure a prominent role for the economic aspects of decision making, we need to further develop and standardise the methods for cost-effectiveness research. The textbooks in which an effort is made to standardise the methodology are especially useful for calculations concerning interventions that focus on middle-aged individuals who suffer from a physical disorder. However, it is more difficult to assess the quality of life of individuals of a young or older age or of individuals who suffer from a mental disorder. Quality of life is typically measured by means of questionnaires on a person's functioning in a number of areas: physically, mentally and socially.

In the case of prenatal preventive tests, too, it is difficult to define an unambiguous measure for the QALYs. As we noted earlier, prenatal diagnostics offers the opportunity for parents to screen their unborn child for several disorders. If it is established that an embryo has a severe congenital or hereditary disorder, parents can decide to terminate the pregnancy. In these cases it is not easy to estimate the number of 'quality-adjusted life years' gained. How can we, in the case of a severe and incurable disease, express the fact that a child will not be born in QALYs? For how many years or generations should we count the costs and effects of children who do not die during pregnancy due to prenatal intervention?

Familial hypercholesterolemia is a genetic disorder characterised by a predisposition to high cholesterol, which often causes individuals to die at a young age from a heart disease. However, by following a strict diet, not smoking and using cholesterol-reducing medication, patients can decrease their risk of dying prematurely because of a heart condition. As of recent, families that have a history of familial hypercholesterolemia can be diagnosed by means of a genetic test, and genetic carriers can start in time with a strict diet and medication. The effects of this intervention will continue to be felt for many generations to come. Here, too, we encounter the question of how to define something in QALYs which affects several generations.

The Social–Psychological Decision: The Vision of the Parents

Looking at the economic assessment of new technologies, we can conclude that there is an increasing interest in rational choice theory models. Therefore, it is an important priority to fill in the blanks in the methodology for establishing the cost-effectiveness of medical interventions. Cost-effectiveness considerations will play an increasingly important role in policy decisions. At the same time, health economists expect that this will lead to more transparent decision making on the social level.

Economic decisions are an important factor in determining which medical interventions will be introduced or covered by health insurance companies. In turn, individuals have a choice whether or not to make use of these interventions. This right of choice is not always invoked consciously; participating is often self-evident. Some issues, however, require a more conscious decision. In the following, I will take the expansion of the Guthrie test as an example in order to examine what considerations play a part in parents' decision-making process regarding new genetic technologies.

The Guthrie test is a blood test that is performed on almost all newborns in most Western countries. Until recently, the drop of blood that was taken from the baby's heel was used to test for 3 diseases, but since 1 Jan 2007, the test has been expanded to a total of 17. The Guthrie test is a form of genetic screening that focuses on prevention of disease. Participation in this test is voluntary, and it thus combines the ideals of preventive medicine—the prevention of disease and suffering—with those of medical ethics—the right to self-determination of individuals. Practice shows that parents consider the first objective to be the most important. From group discussions about the wishes and opinions of parents concerning the expansion of postnatal screening it becomes clear that most parents think the Guthrie test is performed on all newborns. They are unaware of the fact that they can refuse the test. Moreover, while most parents do remember that the Guthrie test was performed, they do not know exactly what it was for. They do not experience this as a problem, however, as they assume that the screening will prevent their child from sustaining permanent damage.

9 Human Enhancement and Genetic Testing

The 17 diseases that newborns are currently being tested for are all treatable. The new techniques now used for the Guthrie test—Tandem Mass Spectrometry—make it possible to detect several untreatable disorders as well. To what extent would parents want to receive information on these disorders? This question was presented to parents in two recent studies performed by Detmar and others. In group discussions, they examined the advantages and disadvantages of the screening for four diseases, which vary in their degree of treatability and moment of manifestation: phenylketonuria (PKU) (see also Chap. 5, this volume), cystic fibrosis (mucoviscidosis), Duchenne muscular dystrophy (DMD) and coeliac disease (gluten intolerance).

All parents support the expansion of screening if it only focuses on treatable diseases and the screening prevents their child from sustaining permanent damage. PKU and to a lesser extent cystic fibrosis are examples of this. If a disorder does not meet these criteria, such as Duchenne muscular dystrophy (a severe and incurable neuromuscular disease) or coeliac disease, parents are far less willing to have their children tested. The reasons they give vary; for example, some express the wish to be able to fully enjoy the carefree time during which their child is not yet visibly ill:

> I would not want to know, I might be burying my head in the sand, but if I knew that within two years my child would be suffering from an incurable disease I would be devastated. I would live with a ticking time bomb in the back of my mind and that would cast a shadow over my entire life.

In addition, knowledge about a child's future development can influence the parent–child bond and the child's upbringing:

> I believe it definitely influences the upbringing of a child, because normally, if your child misbehaves you discipline them. However, if you know that within several years your child will be terminally ill, you might never do so and thus create an out-of-control child.

Some parents indicate that they are wary of the consequences with respect to health insurance, and a few point out that the disease can negatively influence a child's identity development. They do not want their child to be labelled as 'a child with a disorder' even before the appearance of symptoms:

> If you start treatment before a child is truly ill, the child will be stigmatised. This in itself is a form of suffering, because the child will not be able to develop normally.

On the other hand, parents also mention possible advantages of screening, including the opportunity to adjust their lifestyle to their child's illness, a shorter period of diagnostic examination, less uncertainty and choices with regard to future children:

> It is very important to me that I do not have to go through all those months of examinations and tests before knowing what disease my child is suffering from.

> I am the kind of person who wants to know everything, because there might be something I can do, for instance work less. I would also want to know for any future children.

The results of these discussions indicate that the concept of 'enhancement' evokes both positive and negative connotations. They also tell us that it will be complicated to ask for permission to perform the Guthrie test if it also screens for untreatable disorders. Parents have a wide range of ideas on this matter, varying from making the entire screening programme obligatory to keeping it entirely optional. A frequently heard alternative is that a part of the screening programme should be offered in accordance with current practice—treatable disorders only—and to leave the decision concerning disorders that do not meet these criteria up to the parents. Disorders could then be clustered based on specific criteria. For example, a cluster could be formed of diseases that cannot be treated but that can have consequences for the decision to have a second child, such as Duchenne muscular dystrophy. Another cluster could encompass screening that only gives an indication of the chance that a child will develop a certain disorder.

To be able to make a decision, however, a person should have the right information at hand. For participation in a preventive activity, this concerns information on the nature of the disease and the chance of getting it, the possible options for prevention and the possible treatment options. When an individual combines this information with personal values, it is possible to take an informed decision.

Decision aids can assist in this process by providing information that is tailored to the specific needs and situation of each individual. In addition, decision aids can help in weighing the possible options by assigning values to aspects that are of importance to the individual. However, the question remains whether it is possible to satisfy the need for more information that parents experience when deciding on a more complex programme. In particular, the question of whether it is possible for parents to analyse and process this information plays an important role. After all, it does concern a large number of complex medical conditions that are relatively rare.

The possibility to choose seems particularly relevant when screening is offered for disorders for which no suitable treatment is available. Some parents indicate that they do not want their child to be screened for disorders that cannot be treated. Others want to have all knowledge about their child's health as soon as possible. A consequence of offering parents a choice is that the information about the Guthrie test should discuss in detail the effectiveness of early treatment of all the different disorders, which comes with a great risk of overloading the parents with information. Thus, a wider range of options is not necessarily a good thing, especially in the case of choices that are difficult and burdensome. This is all the more problematic because most of these disorders are relatively rare. Offering a wide range of options with regard to the screening programme means that all parents will be asked to make difficult decisions, while only in very few cases an actual disorder will be detected. After all, the chances that your child will suffer from one of the disorders that he or she was or was not screened for are very slim. After the group discussion many participants indicated that they were glad they did not have to make this decision:

> I am very happy I do not really have to decide on this. I would find it an enormous responsibility.

The Social Decision: The Interaction Between Technology and Society

So far, we have discussed the wishes, expectations, experiences and preferences of parents or prospective parents. Parents, as we saw, constitute an important group of consumers of predictive genetic technologies. They have to choose, individually and for each separate pregnancy, whether or not they wish to use these technologies, constantly trying to find their way between the extremes of wanting to know as much as possible on the one hand and not wanting to be unnecessarily anxious or worried on the other. It seems that the parents' choice is the last in a long series of decisions, when new enhancement technologies have found their way into the provision of healthcare. The final question of whether a technology will indeed be used appears to be up to them.

The image of parents, citizens or patients as being the last link in the chain of persons or institutions that decide whether enhancement technologies will or will not be used is based on the presumption that consumers only come into contact with new technologies when these have already been introduced into the healthcare system. Researchers and engineers have dealt with issues of reliability and safety, economists have calculated their QALYs, and now it is up to the individual citizens or patients to decide whether they 'want to know' and whether they will make use of the technological interventions on offer. But is this a correct representation? Is there a linear process of consecutive choices? To what extent is the consumer involved in the earlier stages of the question of how and which innovative technologies will be developed and applied? We will discuss two examples of how consumers—in particular patients—take part in the shaping of healthcare technologies.

Most patients only see the results of scientific and technological research when they visit the doctor's office after experiencing physical or psychological complaints. This is not true for all patients, however. After all, scientific research does not take place in a vacuum, but often includes contact between doctors and patients. Patients and research subjects fulfil a key role in medical research by providing medical data, information about a medical condition, but also research material such as blood, urine and tissue. In some cases, patients go as far as offering their ill body to participate in an experiment, also called a clinical trial.

Participation in a medical study often happens anonymously; patients fill in a questionnaire, give blood or urine and never find out what the doctor did with their samples. In a limited number of cases, especially those concerning rare and severe chronic diseases, the situation is different, and participation in medical research also concerns the patients involved. In the case of rare, hereditary disorders, where doctors are not only interested in the patients themselves but also in their family members, there is often a close relationship between researchers and families or family members. These individuals or patients, who are frequently represented in patients associations, often have a say in the early stages of technology development. This is particularly the case when new developments have been predicted for some time and expectations are running high.

A well-known example is the development of the predictive test for Huntington's disease, a rare and severe hereditary disorder that first manifests itself between the ages of 40 and 50. Parents have a 50 % chance of passing this disease on to their children. Huntington's disease is a neurological disorder that gradually affects the brain and leads to both physical and mental decay. Patients usually die within 10–15 years after the appearance of the first symptoms. In March 1983, researchers localised the gene defect that is responsible for Huntington's disease, thus enabling the development of a genetic test in order to (1) establish whether parents passed the gene defect on to their offspring and to (2) determine whether in turn, this offspring would pass the gene defect on to their yet unborn children. Whereas up until then children of patients with Huntington's lived in uncertainty about their fate—having a 50 % chance of being a carrier themselves—they would now be able to find out in advance whether or not they would get the disease.

The predictive test for Huntington's disease was introduced in the Netherlands 4 years later, in 1987. In the 4 years before the test, researchers examined whether the gene defect that had been localised elsewhere was also present in the Dutch Huntington's population and thus provided information to offer a predictive test. This period was characterised by an intensive cooperation between the patients association—the Dutch Huntington Society—clinical geneticists and researchers. The patients association played an important role in mapping as many Huntington's families as possible. At Leiden University in the Netherlands, a Huntington's archive was created where the blood and data of a large number of Huntington's families were stored. These families hoped to shed some light on the cause of the disease by actively participating in the research.

Not only did the patients collaborate in the scientific research, the patients association was also closely involved in formulating an ethical protocol which would accompany the predictive test. Proper guidance was considered essential by all parties. Individuals who tested positive (meaning that the defective gene is present and that Huntington's will eventually manifest itself) might well sustain a serious psychological injury. If possible, it had to be avoided that this would lead to, for example, depression or even suicide in individuals who knew they would suffer from the disease in the future. In anticipation of these risks, a protocol was developed, stipulating how these prognostic tests should be dealt with. During this process the members of the patients association had an active role. On a number of points they had their own input, which was supplementary to the input of researchers, clinical geneticists, psychologists and genetic counsellors.

The example of Huntington's disease is far from unique; the same is also true for many other patients associations of hereditary and congenital disorders. They, too, cooperate closely with researchers and doctors who treat the disease in question. Researchers are frequent guests at members' meetings and remain in close contact with the members. Not only do they provide information on, for instance, the state of the research, but they also ask for collaboration, for example in the form of research materials and medical data. A member of one of these associations recounts that at almost every meeting some urine was taken from every patient for the purpose of research. Thus, patients are offered hope and the opportunity to help generate new knowledge about their disease.

The American organisation PXE International goes a step further than merely providing research materials and discussing the use of the research results. PXE is a rare genetic disorder of the connective tissue that causes problems at an early age, including blindness, skin lesions and heart attacks. PXE International was founded by Sharon and Patrick Terry, parents of two children who both suffer from PXE. When Sharon and Patrick started gathering more information about the disease and the research that might offer them some hope for the future, they discovered that research on this disease was being conducted at different locations and that the researchers barely shared their results with one another. The Terrys took on what they considered to be an ineffective system, observing that the academic research community is characterised by competition rather than cooperation: as reputation and publications are the cornerstones of a scientist's career, competitiveness often gets in the way of exchanging data and information.

Over the past decades, the Terrys have dedicated themselves to collecting data of other patients, regardless of the rare nature of the disease and the small number of carriers. Thus far, a total of over 50 associations of parents and children with PXE have come into being in many different countries, bringing patients together and mobilising them to provide information and donate bodily materials and fluids. Their blood and tissue bank contains samples from hundreds of individuals. Sharon Terry decided to study molecular sciences and founded a research consortium of 19 laboratories. Not only did Sharon's name grace the patent application when the gene responsible for PXE was isolated in 1999, she also contributed to the development of the first diagnostic test. Accordingly, the Terrys prove that patients and patients associations can have a say in the matter, as well as take matters into their own hands.

Innovative technologies such as preventive genetic tests are often portrayed as being out of our control, whereby developments are presented as a moving train: once the train gets going, it cannot be stopped, and once the train is up to speed, it will only accelerate more. The image of technology as a moving train is a persistent one. As we have shown, however, the train does not just move of its own accord; it is partly driven by prospective users. From an early stage on, these users—in this case, patients—are involved in the development of enhancement technologies, and can thus be called 'co-producers'. New technologies are therefore not only the result of ingenious research and technological developments, but are also formed by the wishes, desires and demands of citizens and patients.

Thus, technology does not follow a linear trajectory from laboratory to prototype to consumer; at almost all stages of this development there is—to some extent or other—an exchange between the scientists' promises about possible applications on the one hand and the formulation of expectations, opinions and definitions of 'good' applications by prospective users on the other.

To Conclude: The Displacement of Politics

The promises of genetic research are substantial, and many agree that it will be possible in the foreseeable future to map an individual's genome at relatively low costs. A person's risk of getting, for instance, various hereditary forms of cancer, diabetes and cardiovascular diseases can be determined by a simple test, and the embryo with the best possible combination of genes can be selected by means of pre-implantation genetic diagnostics (PGD). All these promises are elaborated upon in the movie *Gattaca,* the story of Vincent with which we introduced this chapter. In the world of *Gattaca,* the applications of genetic technology are all-determining, and genetic enhancement has become a reality in many areas of life. The final question we will now try to answer is how likely it is that this scenario will eventually come true.

Discussions on new sciences and technologies and their applications are frequently formulated in terms of moral questions and issues: is it desirable that new technological possibilities radically and permanently transform our society? This type of question presupposes that we—citizens or societies—are in a position to make choices. In *Gattaca,* genetic enhancement is presented as the result of a deliberately chosen scenario.

In practice, however, as we have shown in this chapter, choices about the future are being made in many places at once. The way in and extent to which genetics influences our world and worldviews is determined by a myriad of decisions made by different groups or individuals, at different moments and in different places. Decisions are made in those places where policies are developed and economists discuss the value of QALYs, in the clinics where parents and doctors talk to each other about prenatal and postnatal genetic examination and in research laboratories, where patients donate research materials and medical information in the hope that tests, treatment or medication will also be available to them in the future. Without the QALYs attesting that prevention also has its economic advantages, without parents who are willing to apply genetic tests to their unborn or newborn children and without the commitment of prospective users who are willing to invest in scientific research at an early stage, genetic enhancement would be considerably less self-evident, rendering a society that primarily selects individuals on the basis of their genes rather unrealistic.

Political scientists and science and technology scholars have frequently pointed out that political decisions take shape less and less in political forums. Society in general and the development of science and technology in particular are being shaped to an increasing degree outside of the political sphere, for example in scientific research laboratories, on the free market of consumerism and by the activities of non-governmental organisations (NGOs). At present, the extent to which policy decisions are made in centres of government is decreasing. In political theory, this development is also referred to as the 'displacement of politics'. As we have shown, genetics is a typical example of this development. Choices and decisions are no longer made in the established forums of democracy, but in laboratories, during doctors' consultations, and in the debate among economists. Effective mechanisms capable of steering this 'displaced' power are yet to be developed, according to Noortje Marres (2007). As a result, in our post-industrial and postmodern society science and the development of technology are no longer controlled or coordinated from one central location, but driven by the sum of numerous smaller choices and decisions.

We have been using the movie *Gattaca* as a part of our university Master's programmes for several years now, and with each consecutive year we notice the extent to which our reality is getting closer to the world of *Gattaca*. A recent example is the expansion of PGD to include hereditary forms of breast and colon cancer and the rapidly rising number of Internet-based companies offering genetic tests via the World Wide Web. Every year more tests are carried out, more scientific research into genetic factors is conducted and more patients associations put themselves forward as partners in scientific research. While genetic enhancement may not be an established fact yet, choices are being made on a daily basis that will gradually lead to the routine use of genetic knowledge and technology.

In other words, the reality of *Gattaca* may be more imminent than initially implied at the beginning of this chapter. At the same time, however, it is still a remote future. We are yet to reach the point where every birth is dominated by calculations of genetic probability. Nevertheless, we would like to emphasise that it is of little consequence just how imminent or distant the reality outlined in *Gattaca* is; much more significant is the realisation that we—as citizens, patients or professional experts—can all contribute to the process of societal acceptance and embedding of new science and technologies, more than we are inclined to realise. After all, we can either reject or accept new techniques, by making many small choices, in various locations, day after day.

Bibliography

Asscher E, Koops BJ (2010) The right not to know and preimplantation genetic diagnosis for Huntington's disease. J Med Ethics 36(1):30–33

Detmar S, Dijkstra N, Nijsingh N, Rijnders M, Verweij M, Hosli E (2008) Parental opinions about the expansion of the neonatal screening programme. Community Genet 11:11–17

Detmar S, Hosli E, Dijkstra N, Nijsingh N, Rijnders M, Verweij M (2007) Information and informed consent for neonatal screening: opinions and preferences of parents. Birth 34:238–244

Gezondheidsraad [Health Council of the Netherlands] (1994) Genetische Screening. Gezondheidsraad, The Hague, Publication no. 1994/22

Marres N (2007) The issues deserve more credit: pragmatist contributions to the study of public involvement in controversy. Soc Stud Sci 37:759–780

Nelis A (1998) DNA-diagnostiek in Nederland. Een regime-analyse van de ontwikkeling van de klinische genetica en DNA-diagnostische tests, 1970–1997, diss Twente, Twente University Press, Enschede

Raad voor Volksgezondheid en Zorg [Dutch Council for Public Health and Health Care] (2006) Zinnige en duurzame zorg, Zoetermeer: RVZ, and Zicht op zinnige en duurzame zorg. Zoetermeer: RVZ, available at www.rvz.net

Rutten-van Mölken MPMH, van Busschbach JJ, Rutten FFH (eds) (2000) Van kosten en effecten. Een handleiding voor evaluatiestudies in de gezondheidszorg. Elsevier Gezondheidszorg, Maarssen

Chapter 10
A Unique Copy: The Life and Identity of Clones in Literary Fiction

Bert-Jaap Koops

Abstract Cloning is a typical form of human engineering, which is almost universally outlawed because of ethical objections. But are these objections valid, or are they overly influenced by fictional horror stories? In order to investigate whether clones necessarily lead instrumental lives, have a 'closed future', or lack an identity because they are 'someone else', this chapter discusses fictional accounts of clones. Literary fiction provides a rich picture of clones' lives, demonstrating that clones do not necessarily have to evoke distrust or horror. The mirror that clone fiction holds up to us shows us possible worlds in which a ban on reproductive cloning is not essential to preserve human dignity. Clones may be copies, but they are also unique and original individuals. If we are afraid of cloning, this is not because clones are different or scary but only because society may treat clones inhumanly.

> I knew a little about cloning… but so little that I had not got past carrots, where it all started, to speculate about the notion of duplicating entire higher organisms, such as frogs, donkeys, or people. […] In thinking about this possibility, I found it alarming. I began to see that the duplication of anything complex enough to have personality would involve the whole issue of what personality is—the question of individuality, of identity, of selfhood. Now that question is a hammer that rings the great bells of Love and Death… (Le Guin 1973)

Human cloning seems to be one of the most far-reaching manifestations of the concept of human engineering. Whereas current forms of artificial reproduction,

---

Translated by Mathilde van Heereveld and Berdien Ritzema.

B.-J. Koops (✉)
Tilburg Institute for Law, Technology and Science, Tilburg University,
P.O. Box 90153, 5000 LE, Tilburg, The Netherlands
e-mail: E.J.Koops@uvt.nl
http://www.tilburguniversity.edu/webwijs/show/?uid=e.j.koops

such as IVF, are still dependent on a fertilisation process for which an egg cell and sperm are needed, cloning even skips this last step of the 'natural' process of reproduction. A clone can be made out of any single somatic cell from either a man or a woman, combined with an egg cell from which the nucleus has been removed. By inserting this clone into the uterus, a genetically identical copy of the cell donor can be born. This possibility of cloning that can be used to 'recreate oneself' has amazed and inspired many.

Nevertheless, current technology does not seem ready for this type of cloning yet. Since Dolly the sheep (1997), scientists have succeeded in producing live clones from somatic cells (that is, regular body cells, not egg cells or sperm cells) of several fully grown mammals. However—unless we are to believe the implausible stories of the Italian doctor Severino Antinori or of the Clonaid company that was founded by the Raelian sect—they have not succeeded in doing so with cells of human beings. Nor will this type of cloning be in use soon: not only does the cloning of human beings require experiments that are highly doubtful in the present state of technology, but reproductive cloning is also strictly forbidden in most countries (Brownsword 2008, p. 36). UNESCO's Universal Declaration on the Human Genome and Human Rights (1997) states in Article 11 that reproductive cloning should not be permitted because it violates human dignity. Article 3(2) of the Charter of Fundamental Rights of the European Union forbids reproductive cloning of human beings as well, as it violates a person's right to integrity.

Why is cloning universally disapproved of, and is this a justifiable position? Identical twins are also clones of each other, and society does not object to them. This makes it interesting to further investigate our mistrust or even revulsion of reproducing human beings by means of cloning. Why are we afraid of clones? We do not yet know any clones from real life, and scientific literature can therefore not give us any empirical insights into the life of a clone. This means that we mainly base our image of clones on fictional literature and movies. Images as those from *Brave New World*, *The Boys from Brazil* or *The Invasion of the Body Snatchers* have been engraved into our collective minds. Clones portrayed in fiction are usually no longer humans but products; they do not lead their own lives, but are objects used by megalomaniac individuals or an elitist society. They have no identity because they are actually someone else. Yet that is not the whole story. In other, perhaps lesser-known literary fiction such as *The Cloning of Joanna May* or *Never Let Me Go*, clones are portrayed as normal people leading their own lives. They struggle with the same questions about identity, love and death with which characters in so many other literary works are confronted.

In this chapter, I would like to illustrate the richness of the portrayal of clones in literary fiction, with an emphasis on their life and identity in light of their being clones. How instrumental are their lives in relation to their donors or to society? What does being a clone mean for their sense of identity, and how does society treat them? This journey through the literary landscape will take us through several possible worlds that show how clones do not necessarily have to evoke feelings of distrust or horror. Although the universal ban on cloning renders it quite implausible that one of these possible worlds will become our future world, this does not make the journey any less relevant. On the contrary, the fictive worlds hold up a mirror to us and invite us to reflect on our own, real world and to think about the future we would like to live in.

Cloning and Identity: 'If You are Me, Who Am I?'

In order to understand the descriptions in this chapter, some understanding of cloning and identity is necessary. There are roughly two techniques for cloning human beings: embryo splitting and somatic cell nuclear transfer (SCNT). Embryo splitting or embryo twinning is a primitive form of cloning in which a morula (a zygote that has divided a few times, consisting of 8 or 16 cells—i.e., an embryo at an early stage) is split into two or three parts that continue to grow on their own. This process can be repeated several times with the newly developed parts. Embryo splitting regularly happens in nature as well, for this is how identical twins are formed. This type of cloning can be used to create clones that are genetically identical to each other. However, they will not be identical copies of an already existing person.

This last is the most important difference with SCNT, which is the more advanced and complex of the two cloning techniques. In this technique a cell nucleus, taken from any body cell that is not a sex cell or gamete, is transplanted into an egg cell from which the nucleus has been removed. This egg cell can subsequently develop into an embryo and finally into a neonate, first in a test tube and later in the uterus. The resulting clone will have the genome of the donor, although the copy can never be completely genetically identical, because the mitochondrial DNA (a small amount of DNA that is located outside the nucleus) does not belong to the donor of the body cell but to the donor of the egg cell. As described by Wouters (1998, pp. 39–41), mitochondrial DNA has a minimal influence on a person's makeup. Thus, SCNT does not strictly speaking result in a copy that is completely genetically identical, and can therefore technically speaking not be called cloning. To simplify matters I will leave this distinction aside, as this aspect is neither addressed in fiction nor in the social debate about cloning.

Cloning can have two functions. In *therapeutic cloning*, cloned embryos or cells are used for medical research or therapy. Here, the clones are not implanted and do not grow into human beings. In *reproductive cloning*, the cloned cells are developed further so that they become a 'reproduction' of the donor. Unlike reproductive cloning, therapeutic cloning is allowed in several countries, although sometimes under strict conditions. Since I am interested in questions concerning the lives of clones, I will limit the scope of this discussion to reproductive cloning.

As explained in *Where Idem-Identity meets Ipse-Identity* (Hildebrandt et al. 2008), identity has multiple meanings. Paul Ricoeur made an interesting distinction between idem-identity and ipse-identity. *Idem* refers to being the same. From an external perspective it is determined whether one person is the same as another—for example the same as yesterday, or belonging to a similar group or category. Dr Jekyll is the same person today as he was yesterday and also, in certain ways, the same person as Mr Hyde; his body has continuity in time. Idem-identity thus relates to identification.

Ipse refers to being yourself, as experienced from an internal perspective. Thus, it is concerned with the construction of identity. A person's identity is based on experiences that are used to construct a life story—a process that is stimulated by a sense of self resulting from others' responses to one's self. In other words, ipse-identity is created by interpreting the way others interpret us. This is why Dr Jekyll struggles with his identity: the way in which he is approached as a good, respectable man in social life conflicts with his awareness of his morally bad side, which is taking an increasingly strong hold on him. This is reflected in society's horrified reactions to Mr Hyde's behaviour. While Dr Jekyll shares the same idem-identity with Mr Hyde, he has a split sense of ipse-identity.

A strong interaction exists between idem-identity and ipse-identity. The way others perceive us—for example as Englishman, southerner, school friend, blond or Muslim—influences our sense of identity. Moreover, our ipse-identity influences our behaviour and therefore in turn affects the way others identify us. For clones, this interaction is especially important because the relationship between idem and ipse perfectly illustrates the paradox in the identity of clones. Clones share their donors' idem-identity and *therefore* struggle with their ipse-identity. Being identical makes you struggle with your identity. This paradox is succinctly phrased by Wendy Doniger (1998, p. 136): 'If you are me, who am I?'

Aim and Spoiler Alert

The subject of cloning is present in many works of fiction. However, since the technique has only been known since the second half of the twentieth century, cloning is a less frequent motive or theme than, for example, the Doppelgänger. Aldous Huxley was far ahead of his time when he described the possibilities of cloning in *Brave New World* in 1932. A large number of clone stories were published in the 1970s when the general public, including writers, first began to realise the possibilities of cloning. This can also be seen in the quote by Ursula Le Guin at the beginning of this chapter. In this first period clones could mainly be found in science fiction; other genres followed later, although to a lesser extent, and clones have become widely embraced as literary characters.

In this chapter, I will discuss a selection of novels that include cloning as a key theme. I will limit myself to English literature, where most books concerning cloning can be found. I made a selection of nine novels that cover a wide range of genres: serious literature [Aldous Huxley, *Brave New World* (1932), Fay Weldon, *The Cloning of Joanna May* (1989), Kazuo Ishiguro, *Never Let Me Go* (2005)], suspense literature [Ira Levin, *The Boys from Brazil* (1976)], science fiction [Richard Cowper,

Clone (1972), Kate Wilhelm, *Where Late the Sweet Birds Sang* (1974), Arthur C. Clarke, *Imperial Earth* (1975), Pamela Sargent, *Cloned Lives* (1976)] and young adult literature [Alison Allen-Gray, *Unique* (2004)]. These novels discuss several types of cloning from various perspectives, thus covering a broad spectrum of responses to cloning in literary fiction. I arranged the books according to the different possible functions of cloning: to duplicate extraordinary humans, to reproduce despite infertility, to create a workforce for support and to satisfy scientific curiosity.

Spoiler alert: In this discussion I will give away the plots of several novels. Readers that are unfamiliar with these books (particularly those of Ishiguro and Levin) are urged to read them before reading this chapter.

A Copy of a Unique Original

One of the most important reasons for cloning is to recreate or reproduce a person with unique characteristics. This is not only the case for celebrities with unique characteristics such as Mozart, Gandhi, Einstein or Michael Jordan, who are popular examples of potential clones in the academic literature, but also for persons that one is close to.

In *Unique* (2004) by Alison Allen-Gray, the main character Dominic Gordon discovers that he has an older brother called Nick, whom he has never known. At his grandfather's he finds a photo album containing pictures of someone who looks like himself. However, they depict situations that he has never been in, or where he is older than he is now. He slowly realises that his parents have cloned his older brother, who was a promising scientist, after his tragic early death and that he is the product of this experiment. In his search for what happened a journalist discovers his story, and from that moment on he is pursued by the worldwide press. Dominic is unique because, being a clone, he is the only person in the world that is not unique. The journalist who followed him dies in a cliffhanger scene, and Dominic's story can remain a secret. However, Dominic decides to share his life story with the world, but he combines it with his own message: *all* humans are unique and irreproducible, including clones.

The same motive of cloning a loved one, but this time for very different reasons, can be found in *The Boys from Brazil* (1976) by Ira Levin. After many experiments, Josef Mengele has succeeded in cloning humans as part of his grand plan to recreate his hero, Adolf Hitler: 'his Führer reborn'. He has placed 94 newborn Hitler clones with adoption couples that fit the profile of a young mother and older father employed as a civil servant. Approximately 12 years later the fathers need to be killed, as the death of his father was crucial to Hitler's development. An entire apparatus of killers pays the fathers a visit. It is during these visits that we first encounter the boys from Brazil: arrogant know-it-alls, slightly artistic, with sleek hair. When the original murder plan threatens to fail, Mengele takes the initiative, or rather his Browning, and visits the families himself. When he finally stands face to face with one of the clones he falls to his knees, drooling in worship: 'Mein

Führer!' Failure seems imminent, since the killers have gotten nowhere near the almost 100 clones who, in Mengele's calculations, are necessary for having a good chance of reproducing Hitler. Still, there appears to be some hope for Mengele in the end. The novel ends with a description of one boy who dreamily draws a picture of a large stadium with a charismatic speaker in it, 'sort of like in those old Hitler movies'.

The way in which being a clone influences the clones' lives differs substantially in these novels. The boys from Brazil do not know that they are clones and neither do their parents, who have adopted the boys. Only one of the boys finds out—from Mengele himself, shortly after he has murdered the boy's father: 'You are he, reliving his life!' It remains unclear how the truth about his existence will affect the boy. His last thought about Mengele is that 'he was pretty weird', but at the same time he is very much aware of the power he has over life and death. The novel leaves it up to the reader to imagine what the rest of the boy's life—and the future of the world—will look like.

Dominic from *Unique*, on the other hand, is well aware of the fact that he is a clone. Although this knowledge seems restricting and alienating at first, eventually it gives Dominic the opportunity to reinterpret his own life. All his life his father has been pressuring him to become successful and to develop in a certain direction, but now Dominic understands that he has been living in the shadow of his older brother Nick, whose success he is expected to repeat. However, Dominic proves to be a different person, because apart from genetics, environmental factors also play a role in shaping a person. For example, Dominic is more interested in arts than in science. *Unique* is a Bildungsroman in which Dominic gets to know himself and, in the end, learns to appreciate himself the way he is: 'I'd rather be me than anyone else.' Eventually, he accepts the truth about his being a clone and starts seeing it as a sign of love for Nick and for himself, both of his mother and the doctor who 'engineered' him. Having learned from his experience, he can now continue to live his own life.

Diversity in Unity

The cloning of exceptional persons occurs in two other novels as well. It is interesting to discuss them separately here, as the themes of these novels revolve more around the complex relationship between donor and clone, as well as between different clones, than in the previous novels.

In *Cloned Lives* (1976) by Pamela Sargent, a bioscientist called Hidey Takamura convinces the brilliant astrophysicist Paul Swenson to have himself cloned. Paul simply has too many talents for one lifetime. Takamura wants to be the first person to clone human beings in the new millennium, as soon as the worldwide ban on cloning no longer applies. After a process of prenatal development in artificial wombs, five clones of Paul are born: Ed, Mike, Al, Jim and Kira. The novel subtly avoids mentioning how Kira, a female, was born out of a man's cell—a nice artistic freedom of second-wave feminism, stressing how differences between men and women are caused by nurture rather than nature.

Even though the clones look exactly alike, they develop individual preferences and characters during their youth. Each clone further develops one of Paul's talents, such as writing novels (Jim), mathematics (Ed) and biomedical research (Kira). Despite the fact that the clones grow apart later in life, they have a special connection that sometimes prevents them from having relationships with other people. Also, the outside world treats them as a strange breed of human beings that are a little scary, or even as 'part of the Swenson clone'—thus suggesting that they are one being. Some are more troubled by this than others. Jim is one of the characters who suffer from their heritage: 'He felt he was under an obligation to use his talents for humanity's benefit.'

On the other hand, there are others who confidently follow in Paul's footsteps. Al devotedly continues Paul's work on the moon. When Kira starts to work there too, some sort of family reunion takes place during which the cat is let out of the bag—having been put there by Takamura at the beginning of the story. Paul has been frozen and preserved on the moon, and can be brought back to life because of Kira's efforts. After some starting problems, Paul becomes his old self again and, having been absent for 20 years, continues his life on Earth. In the meantime cloning has gradually become accepted by society, and now humanity faces another fundamental choice: to die, or to continue to a second life in which cloning is no longer necessary because everybody can keep on recreating themselves until the end of time?

The questions with which Joanna May and her clones are confronted in *The Cloning of Joanna May* (1989) by Fay Weldon are a lot more earthly and common: they wrestle with relationships, beauty ideals and ageing. Carl May, a nuclear energy tycoon with cowboy traits and a gigantic ego, had Dr Holly clone his wife without her knowing when she was 30 years old, as she would only get older and Carl rather preferred young women: 'It seemed a pity to let it all go to waste, when you could save it so easily.' The four clones were implanted into the wombs of different women, and grew up separately to become individual persons: Jane, Julie, Gina and Alice.

The novel follows the lives and love stories of all characters during the year that the clones are 30 years old, all struggling with their love lives. The plot twists when the clones meet each other by accident and recognise themselves in one another. They are told by Dr Holly that they are not just twins, but clones. In the meantime, Carl has told Joanna of the cloning during a fight: 'I proved then that you were nothing so particular after all [...] by making more of you, and the more I made of you the less of you there was.'

Paradoxically, however, Carl May has ultimately helped Joanna May become herself by cloning her. After having met her clones, Joanna May no longer feels like Mrs May but refers to herself as 'just Joanna': 'When I acknowledged my sisters, my twins, my clones, my children, when I stood out against Carl May, I found myself.' Jane, Julie, Gina and Alice also learn to live with their new selves in more or less stabilised relationships.

Both novels show how clones develop individually and how their lives differ not only from each other but also from their donor's. The five clones of Paul Swenson and the four of Joanna May have different characters and preferences and lead their own lives. While Joanna's clones only discover that they are clones when they are 30 years old, Paul's clones are aware of their special status their entire lives, all the more so since their environment treats them as clones. Being different marks their lives: 'The others resented us, forcing us together. We had no friends. We sat together, wishing we were like other people.' At the same time, however, they push each other away: 'Oddly enough, their similarities seemed to aid in driving them apart, as if each resented the part of himself he saw reflected in the others.' Their identities are therefore based on their own lives rather than on their common status as clones. Mike, for example, finds the reunion on the moon threatening and dreads 'having to spend time with people who had nothing in common with him except genes. Every meeting and conversation with them threatened his sense of identity.' Because they have strong and idiosyncratic personalities, the clones succeed in developing their own identities and following their own paths.

Jim, the weakest of the group, is the only one who struggles with living in the shadow of his 'father': *'I'm living Paul's life.* [...] He saw himself as a puppet, walking through an ever-repeating cycle.' Jim threatens to commit suicide but his brothers and sister convince him that his life does have value, and he struggles through life as a writer and bohemian. Jim's first novel shows a world full of mirrors and broken glass, thus reflecting his problematic sense of identity: Jim feels fragmented and transparent, wondering who he is when looking in the mirror that are his brothers, sister and father. He still does not know the answer at the end of the novel, but being a writer turns out to be perhaps the most important of all the clones' talents. As a writer he is able to place the technology of the second life in a broader context and to show how this can help in dealing with the human condition. Eventually, he finds his place in the world when Kira tells him, 'You may be the most important of us now, you can write for people, show them how they might realize their dreams. The rest of us don't have much experience with that.'

At the same time, *Cloned Lives* puts the issue of the identity of clones in perspective by portraying a different paradox concerning identity: is a person who is brought back to life after death still the same person? At first the 'man named Paul [...] who sought feebly to imitate Paul's gestures and appropriate his memories' does not resemble the old Paul. As his memories of the past return, he gradually becomes himself again: '*Paul's back*.' However, his memories feel unnatural to him, as '[t]here was no emotional connection with the images of people and far-off places that had settled uneasily into his mind, with the pressured, somewhat frantic individual named Paul Swenson who had existed twenty years before.' He is the same person (idem) but not 'his old self': his ipse-identity keeps evolving in his second life. The crux of these passages is that they emphasise the relativity and dynamics of identity. This is summarised by Kira in a conversation with Jim:

> 'I don't know who's in that room, Kira, but it isn't Paul Swenson. He's not the same person'.
> 'Are you the same person after twenty years? Are you the person he knew before? Think about that. Anyone would be different after so long a time. You're different too'.

This explains why each clone develops his or her own identity and is not bound by the identity of—and the identity with—their donor or fellow clones. A person's self-consciousness (ipse-identity) does not coincide with their genetic construction (idem-identity). It is a combination of genes and environment, nature and nurture, of the clone and the outside world that determines individuality. This combination is unique for every clone.

The Cloning of Joanna May also discusses the formation of identity in human relations, but here cloning plays a different part and is, in fact, a major influence on the identity building of the five women—not because they are treated as clones, but because it holds up a mirror to them and opens their eyes. Through each other they see how they live their lives: relatively dependent and docile in their sexual relations. Together, they learn to take control of their own lives and to be themselves.

The crucial importance of seeing—yourself through the eyes of others—in the development of an individual identity is emphasised in the novel by a word play on the phonetic identity of 'eye' and 'I'. At the start of 'the year of strange events', Joanna reads a story of a girl in Holloway prison that ripped out her own eye, reminding her of Matthew 18:9: 'If thine eye offend thee, pluck it out.' However, this is self-deceit, 'a fine biblical recipe for preserving thy view of thyself as a fine and upright person'. When Joanna hears that she has been cloned she starts wondering who she is: 'The great "I" has fled, say the eyes in the wallpaper: only the clones remain, staring. If the I offend thee pluck it out.' This identity crisis forces her to rediscover herself; all her life she has played the part of 'wife of' but that is not who she truly is. In the course of a year she finds a new identity as an independent woman: 'I, Joanna May. No longer "Eye". Acting; not observing.' The biblical formula for self-deceit can now be reversed: 'I was no longer just a wife; I was a human being: I could see clearly now. If thine eye offend me take a good look at yourself. If thine I offend thee, change it.'

The function of cloning in the novel is to reflect and reinforce Joanna's journey to self-knowledge. Whereas at first having been cloned feels as a loss of identity—'these depletings of my "I"'—eventually the clones strengthen her ipse-identity:

> wife I might be, but only part of me, for all of a sudden there was *more of me* left. The bugles had sounded, reinforcements came racing over the hill; Joanna May was now Alice, Julie, Gina, Jane as well. Absurd but wonderful! [emphasis added]

The clones thus provide Joanna with an opportunity to be herself, more than ever before. The cloning of Joanna May, which started out as a man's trick to keep his wife forever young and subdued, backfires on him. It is a liberating act that frees women and allows them to learn how to be themselves.

The Need for Offspring

Reproducing extraordinary individuals may be a plausible reason to clone human beings, but there is another valid reason: the desire or need to pass on one's genes to the next generation. Infertile humans can resort to adoption or to sperm or egg cell donation, but the wish to have genetically related children can be very strong. Moreover, this kind of reproduction may be the only way for human beings to survive in a society where infertility rates are high.

In *Imperial Earth* (1975) by Arthur C. Clarke, Duncan Makenzie wishes to continue his grandfather's dynasty on Titan, one of Saturn's moons. Since Malcolm was genetically infertile, he decided to have himself cloned on Earth in order to have a son. His son Colin had himself cloned for the exact same reason, and now this second clone, Duncan, in turn travels to planet Earth to create a third-generation clone of Malcolm. During his visit on Earth, he starts to wonder about life and his motivations for continuing the dynasty through cloning: 'Duplication was neither good nor bad; only the goal of it was of importance. And this goal was not supposed to be selfish.' When he eventually brings back a clone to Titan, it turns out not to be a clone of Duncan himself but instead of a talented childhood friend, Karl, whom he had run into on planet Earth and who had died unexpectedly. In the end, Duncan decided that Karl's qualities were more useful for continuing the dynasty under the current circumstances than his own qualities would have been.

A much graver situation features in *Where Late the Sweet Birds Sang* (1974) by Kate Wilhelm. The whole world, including its human population, has been destroyed by an ecological disaster, except for a small community on the United States East Coast that had prepared itself in time for the catastrophe. Most people in the community have become infertile, and the only way to save them from extinction is cloning. This technique is developed by David, a young scientist, and his family. The community rapidly takes shape with new generations of clones. However, the younger generations quickly outshine the older ones, and soon there is no more room for the older, individualistic family members: David is banned from society and the others die out quickly. Group identity becomes the new norm; the groups of clones form unities that can mystically communicate and share feelings even over large distances. Some clones become fertile again, but they try to perfect the technique of cloning in order to make sure that there will be no need to resort to a way of reproducing that is unnatural to them.

Slowly it becomes evident that the later generations are degenerating. They are extremely good at literally reproducing what they have learned, but they have lost all creativity and the ability to think for themselves. Mark, the illegal and naturally born son of first-generation clones Molly and Ben, grows up outside the community and does develop individuality and survival skills. In contrast to the community of clones who eventually prove to be incapable of survival because they lack improvisation skills, Mark survives and is able to create a new community elsewhere. After visiting the old colony, which is destroyed and deserted, he returns home and sees dozens of children full of potential. Mark smiles with happiness '[…] because all the children were different'.

The lives of clones are thematically very different in these novels. In the twenty-second-century world of *Imperial Earth*, cloning is widely accepted, although not common practice. Nobody on Titan is surprised that the Makenzies are clones. Duncan does not experience it as something special; it is sufficiently clear that being someone's genetic copy does not mean having to lead the same life. Malcolm, Colin and Duncan look alike but are different in character, because the subtle differences they had when born were actively stimulated to develop further during their upbringing, so that they would fit best in their time and place. When taking the opportunity to visit the masterpieces of Da Vinci, Picasso and Levinski in the National Gallery of Art in Washington, DC, Duncan becomes aware of the difference between a copy and an original. He recognises the artwork from technically perfect copies, but now he witnesses unique originals. His own decision to break the genetic dynasty underlines that he himself is not only a technically perfect copy but also a unique original.

In the apocalyptic world of *Where Late the Sweet Birds Sang*, clones also differ from their donors. They frequently rebel against the older generation because they feel different, but amongst themselves they are hardly unique individuals. The old generation cannot keep them apart or remember their individual names, and instead provides them with numbers. Every group of around six clones has such a close bond that together they are one being. When one of them experiences pain they can all feel it, and when one of them is in danger in some place the others instinctively know the shortest route there. Their lives are determined by 'the comfort of being brothers and sisters who were as one, with the same thoughts, the same longings, desires, joys'. The novel initially seems to criticise the prevailing individualisation of the Western world in the 1970s; the 'cult of the individual' is a dead end. A clone's sense of self is based entirely on being part of the group: 'We aren't separate, you see. […] If you turned me inside out, there wouldn't be anything at all there.'

Subsequently, however, as the novel follows its main character Mark in rejecting the community as a goal in itself, it gradually becomes clear that it is the lack of individuality that is the real dead end: 'They're all lies! I'm one. I'm an individual! *I am one*!' The loss of creativity and the ability to improvise that comes with the increasing sense of community turns out to be deadly.

That Mark's identity is related to his artistic ability is no coincidence. His mother Molly, banned from the clone society after an expedition that rendered her individualistic, used to make alienating paintings in the same house where Mark now creates clay statues. Through these statues he tries to give meaning to his life and his environment. The importance of an individual creative identity is expressed by Molly in a key passage:

> That other self that speaks to you, it knows what the shape is in the clay. It tells you through your hands, in dreams, in images that no one but you can see. […] Mark, they'll never understand. They can't see the pictures. […] You come here because you can find that self here, just as I could find my other self here. And that's more important than anything they can give you, or take away from you.

Mass Production and the Clone Gap

A third reason for cloning humans in literature is to generate an army of support. This motive can already be found as a subtheme in *Where Late the Sweet Birds Sang*, where a community of clones creates a new generation of clones to carry out labour. The work clones are programmed in such a manner that they accept this without complaining: 'Two castes [...], the leaders, and the workers, who were always expendable. [...] And this would be the final change; none of the new people would ever think of altering anything.' This passage refers to the most famous dystopia in modern literature: *Brave New World* (1932). In this novel Aldous Huxley imagines a world in which five castes are engineered that each have their own tasks. The highest classes, the Alphas and Betas, form the intellectual and executive classes, while the dirty, manual work is carried out by Gammas, Deltas and Epsilons. These lower classes are produced via the so-called 'Bokanovsky process', a simple form of cloning through embryo splitting: 'Making ninety-six human beings grow where only one grew before. Progress.' Everyone is conditioned to be happy with his or her caste and place in society. This is achieved through sophisticated prenatal processes while the clones are being 'bottled' on the production line, as well as in their sleep during infancy. Moreover, the natural development of Deltas and Epsilons is chemically disturbed during the bottling process in order to produce half or complete imbeciles who will not ask stupid questions. The happiness of citizens is ensured by the availability and carefully controlled distribution of the happiness drug soma—'Everybody's happy now.'

This eternal happiness and the subordination of each individual to society are challenged by an outsider, 'the Savage', who was born the old-fashioned way by means of a mother and a womb (dreadful obscenities in the new clinical world) and grew up in a reservation. As a circus attraction, this Savage is guided through a world that is unintelligible to him and which he can only describe in the words of Shakespeare: 'O brave new world that has such people in it.' Eventually, having shocked the higher castes too much with his utterances and behaviour, the Savage is put away in a hut where he hopes to live his life in seclusion. However, he is soon discovered by tourists who pressure him to join in a group orgy that causes his downfall.

A different gap in society with clones as the lower caste can be found in *Never Let Me Go* by Kazuo Ishiguro (2005). This novel tells the story of Kathy, who at 32 looks back on her nearly completed life. She talks about how she grew up at Hailsham, a closed, protected and elitist institution, and the triangle that developed between her, Tommy and Ruth. Similar to the students, the reader only gradually discovers what they are and what kind of world they live in: they are clones, predestined to donate their organs. After leaving school they start working as carers for older donors and eventually become donors themselves. Unless complications occur earlier, they will have fulfilled their life's purpose with the fourth donation when they 'complete'—probably a euphemism for dying.

Tommy and Ruth have an on-and-off relationship. Even though Kathy and Tommy would be a better match, they never managed to start a relationship. After

10 A Unique Copy: The Life and Identity of Clones in Literary Fiction

Ruth 'completes' with her second donation, however, Kathy becomes Tommy's carer, and they can give their love free reign. They hear a rumour that when two clones love each other enough they can receive a stay of donations. The rumour, however, is false, like many of the myths that circulated among Hailsham students. Miss Emily, the former school principal, tells them how grateful they should be for having grown up at Hailsham: 'Look at you both now! You've had good lives, you're educated and cultured.'

Hailsham turns out to have been an attempt of Miss Emily and others to show the world that clones are humans too, by giving them a proper education and displaying their art: the mirror of the soul. However, the climate changed and Hailsham had to be closed. Society does not want to return to a cloneless world full of illness, but neither can they face the reality behind the system of organ donations. The clones—'[s]hadowy objects in test tubes'—are now again hidden away in unknown places under wretched conditions. This is possible because they are not 'like us'; they are 'less than human'.

Both *Brave New World* and *Never Let Me Go* describe a lower class of clones that consists of humans whose lives are instrumental in relation to the ruling class. However, the books are very different and express diverse meanings. *Brave New World* is not so much about cloning or human engineering as it is about social engineering. It is a political novel showing the consequences of a totalitarian society that has perfect control over economic production processes. The clones exist to serve society, and all of them, including the Alphas and Betas, are conditioned to consume as much as possible in order to keep the economy running. Here planned economy translates into planned life: 'People are happy; they get what they want, and they never want what they can't get. [...] [T]hey're so conditioned that they practically can't help behaving as they ought to behave.' Clones who start thinking about their lives and realise that the ultimate goal in life is not happiness or pleasure but rather sharpening the mind and gaining knowledge are seen as a threat to the state and banned to an isolated island.

Huxley shows the problematic consequences of utilitarianism and simultaneously argues against state ideology and totalitarian regimes. His use of reversal as a major stylistic technique is very effective, as exemplified by the word 'mother' being an insult and Shakespeare's works being seen as uncivilised and full of nonsense. Thus, Huxley shows the consequences of society's constant strive for perfection. This underlines the message from the book's motto: 'Perhaps a new century will start; a century in which intellectuals and the cultivated classes will dream of ways to avoid utopias, to go back to a society that is not utopian, less "perfect" and more free.' Freedom also means making one's own decisions, including the choice to *be able to* be unhappy (cf. Chap. 12, this volume):

> 'But I don't want comfort. I want God, I want poetry, I want real danger, I want freedom, I want goodness. I want sin'.
> 'In fact,' said Mustapha Mond, 'you're claiming the right to be unhappy'.
> 'All right then,' said the Savage defiantly, 'I'm claiming the right to be unhappy'.

Not to mention the right to grow old and ugly and impotent; the right to have syphilis and cancer; the right to have too little to eat; [...] the right to be tortured by unspeakable pains of every kind. There was a long silence.
'I claim them all,' said the Savage at last.

At the same time, *Brave New World* has some interesting things to say about the life and identity of clones, even though they—literally and figuratively—play a subordinate role in the novel. Society is appalled by the concept of identity. The idea is that everybody loses their sense of identity by dissolving into the community with the help of soma and group orgies. Everybody who is 'somebody', with their own ideas, gets banned. Clones within a Bokanovsky group have no identity apart from a sense of community that gives them the idem-identity of being a group member. Not only do the clones lack an individual ipse-identity, they can also be seen as repulsive as a group. They are described as swarms of insects:

Twin after twin, twin after twin, they came—a nightmare. Their faces, their repeated face—for there was only one between the lot of them—puggishly stared, all nostrils and pale goggling eyes. [...] In a moment, it seems, the ward was maggoty with them. They swarmed between the beds, clambered over, crawled under [...].

The long rows of 'identical midgets', the 'twin-herds', the 'human maggots' and 'lice' form a 'nightmare of swarming indistinguishable sameness'. They exemplify the terrible absence of individuality in a *Brave New World* in which art has made place for shallow 'feelies'–multi-sensory films—for the masses. In the same way that clones connect with their group, every individual connects only with the here and now. The humans, similar to the 'feelies' they experience, only give meaning to the pleasures of the moment, which makes it impossible for them to construct a life story with a history and a future, and consequently to create a unique identity. Just like the clones swarm about in 'indistinguishable sameness', the totalitarian planned economy and the planned life swarm about like a plague of grasshoppers, ensuring that nothing remains of the meaning that could have been attributed to life and being human.

Clones are experienced as scary creatures in *Never Let Me Go* as well. Even though she is concerned with their fate, Madame—one of the school's patrons—shrinks back in fright from the clones: '[...] she saw and decided in a second *what we were*, because you could see her stiffen—as if a pair of large spiders was set to crawl towards her.' A crucial difference with *Brave New World* is that there the clones are seen as sordid inferiors by both the Savage and the reader, which makes the clones seem despicable and creepy. In *Never Let Me Go* the reader sees through the eyes of the clones and gets to know them as ordinary persons. They experience the same developments and feelings, love and sorrow that are part of ordinary human life.

This novel addresses two interrelated ethical objections that are frequently discussed in the academic literature about cloning: the treatment of human life as instrumental and the 'closed future' of clones, which we also encountered in *Unique*. Although the clones of Hailsham maintain the illusion that they can lead their own lives, they simultaneously and often subconsciously realise that their future is already fixed.

This is underlined by a stylistic technique that is applied throughout the novel. Kathy tells her life story in such a way that the reader feels the story has already been told before: it is as if Kathy and the reader are trying to remember the story together. Episodes are often introduced with demonstratives such as 'that', as in the sentence 'what happened *that day* at the pavilion when we were sheltering from the downpour' (emphasis added). Kathy also uses the phrase 'of course' rather abundantly. For example, when describing her search for a song in second-hand shops in Norfolk—Judy Bridgewater's 'Never Let Me Go', a song that meant a lot to Kathy when she was young—she remarks: 'Then of course I found it'. Through this technique, the novel subtly suggests that the future is fixed and that everything had to happen the way it did. Perhaps this is the main theme of Ishiguro's fiction: the realisation that time cannot be reversed, and that choices once made, however sensible they seemed at the time, will have consequences for the rest of your life.

The lives of the clones in *Never Let Me Go* are ambiguous, for even though their future is fixed, the clones can lead full human lives by making choices about the things that matter to them. Possibly, within the limits of their destiny of becoming donors, Kathy and Tommy could have led very different lives, but they realise this only afterwards. Ishiguro's fiction demonstrates that, at the end of the day, all people have a 'closed future'. The freedom to make choices—demanded so passionately by the Savage in *Brave New World*—has two sides to it. What are the crucial choices in life is usually not discovered until long after those choices were made, when it is too late to reverse the choice or to change one's mind. Thus, *Never Let Me Go* suggests how, despite their closed future, the lives of clones are not fundamentally different from those of other humans.

A similar ambiguity can be found in Hailsham. The clones play an instrumental part in the organ donation programme and as such form a lower class, but the students of Hailsham have a privileged position among the clones; they are the elite of the lower class. On the one hand, this is a place that brings happiness to the students; the name can be read as Hails-ham, a home or place of hails, referring to the old English use of 'hail' as 'health' or 'well-being'. On the other hand, it is a place where the clones are being fooled into thinking they have health or well-being: it is a Hail-sham. In both senses, Kathy's identity is largely based on her position as a Hailsham student.

The clones base their identity mainly on the role they are given and the group connected to that role. At school the clones' identity is determined by the different roles they take on as students, at the Cottages by whether or not they went to Hailsham, and later in life by being a carer or organ donor. The first time the clones realise they are different is when, at a young age, they challenge Madam and see how she trembles with revulsion:

The first time you glimpse yourself through the eyes of a person like that, it's a cold moment. It's like walking past a mirror you've walked past every day of your life, and suddenly it shows you something else, something troubling and strange.

When they grow older they try to find comfort in being a clone by searching for 'possibles', persons that could be their genetic donors: 'We all of us, to varying degrees, believed that when you saw the person you were copied from, you'd get *some* insight into who you were deep down, and maybe too, you'd see something of what your life held in store.' Kathy looks for 'possibles' in porn magazines because she sometimes has strong sexual desires and reasons: 'It has to come from somewhere. It must be to do with the way I am. [...] So I thought if I find her picture, in one of those magazines, it'll at least explain it. I wouldn't want to go and find her or anything. It would just, you know, kind of explain why I am the way I am.'

Even though their identity is partly determined by *what* they are—being treated differently as clones—the question of *who* they are is at least equally important for the clones' identity. This is vaguely filled in by the possible genetic donors, but the 'possibles' are another Hailsham myth that the clones only partly believe in. In the end, their identity is determined, like that of 'normal' humans, by their daily contacts with friends, classmates, companions and loved ones, all of whom will from time to time hold up a mirror in which they will see something 'troubling and strange': themselves through the eyes of others.

The Curious Scientist

Sometimes there is no intrinsic reason to clone except for curiosity: it is interesting to investigate whether cloning is possible and, if so, how it works. In *Unique* and *Cloned Lives* we already encountered curious scientists who were the driving force behind cloning. This can be seen in *Unique* when Professor Imogen Holt explains to Dominic why she made him: 'I wanted to see if it *could* be done.' While the doctors in these novels try to apply cloning to a useful purpose, Professor Miriam Pointer in *Clone* by Richard Cowper (1972) has no such intentions:

'When you first thought of making them, Miriam, did you have any idea what you were doing?—apart from breaking the law, I mean'.
'No, not really,' admitted the Professor. 'It just seemed a rather fascinating piece of research'.

Pointer's main goal was to investigate whether she was able to produce a child with a fully eidetic memory from two special parents. The four clones, Alvin, Bruce, Colin and Desmond were cloned through the simple technique of embryo splitting. As with the sorcerer's apprentice, however, this fascinating piece of science becomes bigger than she can handle. When the clones are 15 they meet each other and discover they have supernatural powers. They change Pointer's head into funny animal shapes and teleport her naked to the hallway. In an extreme

10 A Unique Copy: The Life and Identity of Clones in Literary Fiction

response to these boyish pranks she uses strong chemicals to erase their memories and with that their identities.

It is 3 years later when the main character, Alvin, regains his memory: 'His lost identity streamed back into his consciousness like sand in a twisted hour-glass. [...] "I am Alvin Forster, an eidetic freak. *And there are four of me.*"' He reunites with his brothers, and together they are able to take on the entire world: 'It wasn't just a case of 4 times 1, but of 1 to the power of 4! Or maybe even 4 to the power of 4!' They are not human, but 'a supra-human species of virtually uncalculable powers'. Remarkably, they are not a scary species that threatens humanity. Alvin and his brothers are pure, innocent creatures with 'qualities of saintliness'. However, humanity is not ready for them, and instead of saving the world they retreat to a parallel world.

The lives of the clones in *Clone* are closely related to their sense of identity. The main part of their lives they are unaware of each other's existence or of the fact that they are clones. They live normal, somewhat boring lives, especially during the period they have lost their memories. After they are reunited and rediscover their combined strength—they are able, for example, to communicate with their thoughts and manipulate things at a distance—their sense of identity changes radically:

> He closed his eyes and opened them again upon Desmond and Colin and Bruce, who were but Alvin and Alvin and Alvin. Four to the power of four. But four what? 'Clones' she had called them. 'I am we,' he murmured, 'we are I.'

The four clones increasingly become a unity of four and eventually manifest themselves as one person. Seamus O'Duffy wants to destroy the clones by order of the European chiefs of state because they are seen as a threat to humanity. When he talks to the clone through a video connection and asks, 'And who might you be, sir?', the clone blinks his eyes for a moment:

> 'Well, do you know,' he said, 'I've never really thought about it.' Then his brow cleared. He smiled. 'My real name could well be Adam,' he said. 'Adam Clone.'

The merging into one newborn—or newly engineered—post-human also has a practical reason. All four are in love with Cheryl, their saving angel who accompanied them during all of their adventures. Now they no longer have to compete with each other or be distracted by each other. Cheryl gets four for the price of one.

This witticism, very fitting for the hippie age in which the novel was written, is typical for the humoristic tone of the novel. *Clone* offers a light and casual perspective on the future possibility of human cloning, which makes it different from most of the other novels discussed above. It is also exceptional in clone fiction to plainly portray clones in such a positive light; Cowper's clones take a great leap forwards in the evolution of humankind. However, this message is buried beneath a layer of humour and slapstick, and the novel does not want to be taken seriously. According to Cowper, clones are nothing but 'a rather fascinating piece of fiction'.

Conclusion

Our tour through the lives of clones in literary fiction has provided us with a wide variety of images. While in some literary fiction the worries about the instrumental lives and identity crises of clones that contributed to the worldwide ban on cloning are confirmed, we have also seen examples of fictional worlds in which the cloning of human beings is quite compatible with human dignity. In some stories clones are depicted as frightening creatures: as insects (Huxley), as part of the doom scenario of Hitler's resurrection (Levin) and as alien creatures with mystical powers that lack individuality (Cowper, Wilhelm). In several other stories, however, the reader gets to know clones as ordinary people with everyday problems (Clarke, Weldon), even if they are regarded as frightening by the societies they live in (Allen-Gray, Sargent, Ishiguro).

Upon closer examination, the literary works discussed create a fairly nuanced image of the life and identity of clones. The assumed instrumentality of clones emerges as an important theme, especially in cases where the clone is a copy of an exceptional person or has come into existence to save the human species from dying out. For some, such as Dominic (*Unique*) and Jim (*Cloned Lives*), this leads to a 'closed future'. They live their lives in the shadow of their instrumentality and are unable to determine their own path in life.

However, the image of a closed future generally turns out not to be true. Dominic mainly suffers from living in his brother-donor's shadow because his father raises him as if he were his brother. Eventually, he manages to break free from his father's expectations and is then able to live his life the way he wants, together with his mother. The brothers and sister of Jim, too, show us that the donor does not necessarily have to be an oppressive shadow; he can also be a shining example to his clone children. The notion that a clone's future becomes fixed once they encounter their donor—'so this is me in thirty years'—can also be reversed. This knowledge, as suggested by Julie in *The Cloning of Joanna May* ('you might learn something from yourself grown old'), can also be used to make better-informed choices about how to live your life, taking your possible future into account. Whether somebody would experience life as a clone as a burden restricting personal freedom is therefore questionable: in the novels, this only happens when clones with an introverted personality are raised under pressure to fulfil a particular destiny. Although the clones in *Never Let Me Go* lead instrumental lives and are part of an organ donation programme, they are able to lead their own lives within this context. For Kathy, her own choices—or lack of choices—about her relationships are what matters in life. The tragedy of the missed relationship with Tommy is greater than the tragedy of being an organ donor. But the most influential choices only become clear when it is already too late. Also in this sense, clones are just like regular human beings. At the end of the day, the future appears to be closed for everyone.

Ishiguro's fairly deterministic attitude towards life is not shared by Huxley or any of the other writers that champion individual autonomy. Freedom of choice,

10 A Unique Copy: The Life and Identity of Clones in Literary Fiction

including the Savage's choice to be unhappy, is necessary to shape one's own, individual identity. Life as a clone threatens to minimise individuality—as Jim observes, the clones are 'denied even the small pleasure of feeling like unique individuals'. Nevertheless, on average the novels seem to refute this negative view of clones' identity building. At first, Joanna May experiences an identity crisis when she hears she was cloned ('if the I offend thee pluck it out'), but she is soon able to convert this feeling into the positive feeling of a new and more authentic identity ('if thine I offend thee, change it').

Dominic goes through the same phase when he realises that every human being is unique, including himself as a clone. For Kathy and her friends, their upbringing at Hailsham provides them with the capacity to become complete human beings, independent from their 'possibles', their donors. Several novels underline that identity building is a dynamic process in which memories play an important part. Memories are, in fact, the stories of our lives: without memories, Paul in *Cloned Lives* and Alvin in *Clone* would have been nobody, or at least not 'themselves'. The continuous stream of events in a human life has at least as much influence on the sense of self as genes or upbringing. Clones do not differ from other human beings in this respect either. It is as Kira remarks: 'Anyone would be different after so long a time.'

Many novels thus show that clones are perfectly capable of developing and shaping their own identities, but there are also novels that convey a different image. In some novels, clones form groups and possess a strong collective—rather than individual—sense of identity. However, this only occurs in the novels of Wilhelm and Cowper, where clones share a mystical and telepathic bond. In these books, cloning fundamentally changes something in their being, and as a consequence the clones identify with the group instead of with themselves. The clones in these books are no longer human beings. In several other stories a group identity is present as well, but this originates from other causes. In *Brave New World*, group identity is created through technical and social conditioning, while in *Cloned Lives* and *Never Let Me Go* society helps develop group identity by treating the clones as groups.

These last novels show that clones are again not very different from 'normal' humans. The clones from the 'Swenson group' develop individual identities mainly because they recognise themselves in their group members and want to be different, just like Kathy and her classmates at Hailsham. Group bonding is a human trait and a social necessity that helps form identity. From time to time, a mirror is held up to us when interacting with friends, peers and loved ones. In a way, for clones this mirror shows a double reflection: they can directly recognise themselves in the other person as well as see themselves through the eyes of that person. This can complicate their identity, but it can also give them an opportunity to construct a stronger individuality for themselves.

Altogether, it becomes clear from the life and identity of clones in literary fiction that cloning does not necessarily violate human dignity or personal integrity. The social fear of cloning appears to be rooted in spectres from fiction, such as Mengele's boys from Brazil, the human maggots from *Brave New World* and the mystic clones in the novels by Wilhelm and Cowper. The last two seem a lot less threatening compared to the novels of Levin and Huxley, as the mystic and telepathic clones of Wilhelm and Cowper are clearly fictitious and do not pretend to be real. While Levin's doom scenario and Huxley's dystopia are also unrealistic, they do provide a realistic warning for what might happen if reproductive cloning were to be used on a larger scale to redevelop society in a way that deviates from constitutional democracy. However, the novels also show that the danger does not lie in the clones' genome but in the way they are conditioned. It is authoritarian fathers and totalitarian life that make clones what they are. In addition, the social divide portrayed by Ishiguro is not a dystopia because of the development of clones, but because society has gone too far in its urge to engineer health. As explained to Kathy by Madame, biotechnology has driven humanity out of society:

> I saw a new world coming rapidly. More scientific, efficient, yes. More cures for the old sicknesses. Very good. But a harsh, cruel world. And I saw a little girl, her eyes tightly closed, holding to her breast the old kind world, one that she knew in her heart could not remain, and she was holding it and pleading, never to let her go. That is what I saw [...] and it broke my heart.

Anyone travelling through the landscape of clone fiction can see that, instead of horror images, the novels generally give a nuanced image of clones. The mirror that clone fiction holds up to us shows us possible worlds in which a ban on reproductive cloning is not essential to preserve human dignity. Genes only tell a small part of the story. Clones may be copies, but they are as unique and original as their donor. If we are afraid of cloning, this is not because clones are different or scary but because society may treat clones inhumanly. Thus we come full circle: clones are different only because they are seen and treated as different.

This circle can be broken, as shown by our fictitious heroes Dominic, Kira, Duncan, Kathy and the clones of Joanna May. Anyone who treats them as humans gives them the freedom to live their own lives and construct their own identities. Yes, they are different because they are copies, but they are also unique because they differ as individuals from their donor—just as every human being is unique, cloned or non-cloned, enhanced or non-enhanced. Are we not all different?

Bibliography

Fiction (Edition Used Between Brackets)

Allen-Gray A (2004) Unique. Oxford University Press, Oxford
Brownsword R (2008) Rights, Regulation, and the Technological Revolution. Oxford UP, Oxford
Clarke AC (1975) Imperial earth (Gollancz, London 2001)
Cowper R (1972) Clone. Doubleday and Company, Garden City
Huxley A (1932) Brave new world. Chatto and Windus, London (Harper and Row, New York 1969)
Ishiguro K (2005) Never let me go. Faber and Faber, London

Levin I (1976) The boys from Brazil. Michael Joseph (Pan Books, London 1977)
Sargent P (1976) Cloned lives. Fawcett Publications, Greenwich
Weldon F (1989) The cloning of Joanna May. Collins, London (Flamingo, London 1993)
Wilhelm K (1974) Where late the sweet birds sang. Harper and Row, New York etc. (Gollancz, London 2006)

Secondary Literature

Doniger W (1998) Sex and the mythological clone. In: Nussbaum MC, Sunstein CR (eds) Clones and clones. W.W. Norton and Company, New York, pp 114–138
Hildebrandt M, Koops BJ, de Vries K (eds) (2008) D7.14 a: Where idem-identity meets ipse-identity. Conceptual explorations. FIDIS Deliverable. Available on www.fidis.net/
Le Guin U (1973) On theme. In: Robin SW (ed) Those who can: a science fiction reader. Mentor, New York, pp 203–208
Salgueiro Seabra Ferreira MA (2005) I am the other. Literary negotiations of human cloning. Praeger, Westport
Wouters P (1998) Een gekloneerde toekomst. In: Henriëtte B (ed) Allemaal klonen. Feiten, meningen en vragen over kloneren. Boom/Rathenau Institute, Amsterdam, pp 19–43

Chapter 11
Parents' Responsibility for Their Choices Regarding the Enhancement of Their Child

Carla Sieburgh

Abstract This chapter explores the question whether a child can hold his or her parents legally liable for damage caused by a decision made either during the pregnancy or after the child was born. Can a child demand to be paid a compensation for the damage that their decision has caused? The answer, in short, is 'no'. In order to receive compensation the child has to prove that the decision made was faulty, which appears to be difficult, for parents have the legal right and the freedom to make a decision based on their beliefs of what is right and wrong. Medical caregivers, on the other hand, are liable for damages caused by their not supplying the parents with full information on risks and choices incurred during and immediately after pregnancy.

> At the point where the difference between 'being' and 'being available' becomes blurred, all aspects of human life—health, happiness, conception, knowledge, degrees—present themselves as human rights. And when the supply of universal usability spreads all over the world, not even the citizens themselves—indeed, in particular not the citizens themselves—feel responsible for the world any longer. Instead, they angrily and inexhaustibly pursue claims of damage. (Alain Finkielkraut, 'Créanciers du monde', in: *L'imparfait du présent,* Gallimard 2002, p. 278)

John wants to get rid of his glasses. He does not think they suit him and believes contact lenses may suit him better. Therefore, he decides to buy contact lenses at the drugstore, choosing a lens power similar to that of his glasses. Soon after John starts wearing them, however, his left eye becomes infected. The antibiotics take

---

Translated by Laura Jannink.

C. Sieburgh (✉)
Radboud University Nijmegen, Faculty of Law, P.O. Box 9049, 6500 KK, Nijmegen, The Netherlands
e-mail: c.sieburgh@jur.ru.nl

effect, but after the treatment John notes that his sight in the affected eye has deteriorated significantly, leading him to seek damages from the company who sold him the lenses. Against the background of the general topic of human enhancement, it becomes clear that John used a particular instrument—contact lenses—to alter or improve his life or body. In this example, John makes his own decision regarding the enhancement of his body, and it is he who experiences the positive and negative consequences of this decision.

To some extent, human beings can decide on the ways and means they wish to use to organise their lives according to their own insights, and any disadvantageous consequences that are intrinsically linked to those decisions are accepted. In the case of disadvantages that customers could not have anticipated, or that can no longer be considered reasonable, the manufacturer or seller can be held responsible for the damages their product has caused. Conversely, if somebody decides not to make use of the available instruments, he or she is personally responsible for any detrimental consequences resulting from this choice. The present chapter, however, does not concern those cases where the person doing the modifying is also the one being modified—the enhanced human being.

The question of liability for a particular choice becomes more interesting when the person who applies the instrument is not the same person as the one to whom the instrument is applied. Imagine, for example, a cardiologist who instead of performing an angioplasty to treat a patient suffering from atherosclerosis decides to use a stent or to issue a referral to a cardiac surgeon for bypass surgery. Or think of parents who decide that their child should not wear braces or who control their child's destiny by means of prenatal screening (ultrasound scan, blood test, amniotic fluid puncture—compare Chap. 9, this volume), abortion or vaccination. In these latter cases, we are dealing with parents who make a decision, instruments that may or may not be used and children who will experience the positive or negative effects of their parents' decisions. Nothing happens if the choice turns out to be the 'right' one, but if something goes wrong the child will suffer the consequences. The manufacturer of the instrument, the attending physician or the health insurer may then face a liability suit for damages caused by their product or actions. In addition, the question arises whether children can hold their parents legally liable for any harmful effects resulting from their decisions.

There is no fundamental difference between a decision regarding braces and a decision regarding inoculation. Nevertheless, public opinion appears to differentiate between the two. This might be accounted for by the fact that inoculation usually concerns babies, who are more appealing to the emotional imagination than pimply teenagers in need of braces. Still, it is remarkable that questions regarding parents' liability for their choice of whether or not to inoculate tend to evoke a certain indignation, whereas liability for the choice regarding braces is generally considered from a more rational point of view.

There *are* fundamental differences, however, between the decision to terminate a pregnancy and the decision to inoculate; the existence or non-existence of a child is of a wholly different order than the vaccination of an already existing child.

11 Parents' Responsibility for Their Choices Regarding the Enhancement 153

Nevertheless, the idea of liability for both types of choices evokes a similar response of shock or disbelief.

What makes these choices special? It seems surprising that a society which tends to think in terms of 'formats' and protocols, grouping together different variations on a theme, should hold such fundamentally different views on inoculation and braces.

In this chapter, I seek to explore the difficulties in answering the question of whether children should be able to hold their parents legally liable for damages caused by their choices during or immediately after pregnancy. In the following sections, I will explain that liability laws describe a number of requirements which must be met before liability to pay damages applies. Furthermore, it will be shown that the literal application of liability laws can lead to outcomes which are at odds with explicit or implicit assumptions and feelings held by our society, as embodied in the response of shock or disbelief mentioned above. The fact that parents cannot yet be held responsible for such choices does not follow directly from the existing liability laws. But what, then, is the cause of this situation?

Examples

In May 2010, Evelyn gave birth to a baby boy named Timmy. According to the normal vaccination schedule Timmy is due for his first set of vaccinations around his first birthday in the summer of 2011, in order to immunise him against bacterial meningitis. His parents, Paul and Evelyn, have done a great deal of reading on the possible harmful effects this vaccine can have on the nervous system of a child his age. Hence, they decide to postpone the vaccination for a year, but in the fall of 2011 Timmy contracts bacterial meningitis. Moreover, soon after he recovers from the infection it becomes evident that his motor development has been affected.

Question: is it possible for Timmy—when he is older and realises he is handicapped—to hold his parents liable for the damage their decision has caused? And what exactly does this damage entail? When considering legal action, Timmy will base his claim on the idea that the government must have put a lot of thought into the vaccination schedule, and that while the programme is not risk-free, these risks were obviously weighed against the risks of postponing or omitting vaccination and found to be the lesser of two evils.

This case illustrates how human enhancement gives rise to the question of whether the persons choosing enhancement—or non-enhancement—should be considered responsible for their actions. As will be explained later, this question is even more difficult to answer when specifically applied to parents making a choice regarding their child. Dilemmas arise immediately after a child is conceived.

Consider the following situation. In light of Evelyn's age (40), Paul and Evelyn have to decide on getting an amniotic fluid test in order to establish whether their child has a congenital defect. They are aware of the fact that the likelihood of such a handicap becomes greater with increasing age of the mother, but they also know that the amniotic fluid test is not without risk. What if it results in miscarriage, while the child turns out to have been perfectly healthy? Hence, Paul and Evelyn decide against this and other, less risky, procedures. After Evelyn gives birth, however, they discover that their daughter

> Jasmine has a cleft palate. While this defect is operable, Jasmine will always look slightly different and will also have severe speech defects. At times, Jasmine is greatly troubled by these problems, and Paul and Evelyn wonder if they made the wrong decision.

What is the current situation with respect to liability for harmful consequences that follow from the possibilities regarding human enhancement? While there may be an interconnection between the decision to act or to refrain from acting (the choice between enhancement and non-enhancement), the responsibility for this decision and the provisions of civil law concerning liability for harmful consequences, in the following I will argue that the one does not follow logically or automatically from the other.

Legal Questions Concerning Human Enhancement

Human enhancement is increasingly becoming a part of life, starting even before conception. Women who want to become pregnant are advised to take folic acid to decrease the risk of the baby being born with *spina bifida* (literally: 'split spine'). During pregnancy, parents are confronted with a range of options concerning prenatal diagnostics, including ultrasound scans, a nuchal scan that may be combined with a maternal blood test, a chorionic villus sampling (CVS) and an amniotic fluid test. These options are available to all parents, regardless of whether they have reason to believe their child may have a congenital defect—as would for example be the case if there is a family history of an inherited condition. After the child is born, further choices have to be made regarding inoculation or screening.

The options and possibilities have become more sophisticated over the years. Until the end of the eighteenth century it was not legally possible to terminate a pregnancy. Women who wanted to have an abortion had to turn to an illegal surgeon or poisoner who would try to kill the foetus using a heated knitting needle. Moreover, the decision to abort was not based on knowledge regarding the baby's health or possible handicaps, as prenatal diagnostics did not exist at that time. Even after birth the possibilities to determine or influence a child's state of health were limited, and many children died of infections—which can now be treated and cured with penicillin or other antibiotics—due to a lack of hygiene.

Clearly, a lot has changed since then. Today, it is possible to legally terminate a pregnancy—within certain limits laid down by the law. Another modern option is to perform prenatal diagnostics in order to acquire some insight into a child's state of health, on which parents can then base their decision to either continue or terminate the pregnancy. If the results reveal a congenital defect, parents are not obliged to choose abortion.

Human enhancement implies making choices. Whenever enhancement is on offer, humans have to consider the possibilities it provides and the consequences that follow from either embracing or rejecting the different options.

Making an irreversible choice naturally precludes following the opposite route. If prenatal diagnostics establish that a child has a congenital defect and the parents still decide not to terminate the pregnancy, the child will be born with a handicap. This may come with disadvantages that could have been prevented by opting for abortion. Suffering or reliance on assistive devices, for example, might cause a child to feel that it would have been better if he or she had never existed.

Whenever such disadvantages follow from an essentially legal choice made by the parents, several questions arise regarding their possible liability. Parents are responsible for their decisions, but does this also imply that they can be held liable if a choice turns out to be harmful? Can a child claim damages for a decision that he or she considers to be the wrong one? Moreover, what about a child who is born disabled after the parents consciously decided not to perform prenatal diagnostics? The dilemma becomes even greater when the results of such diagnostics reveal that a child has a congenital defect and the parents still decide not to terminate the pregnancy.

Questions may also arise regarding choices made *after* birth. For example, can children who contract meningitis as a result of their parents' decision not to inoculate hold them liable for the damage suffered from this decision? Or consider the reverse situation: children who suffer delayed physical and mental development as a consequence of vaccination. Can they hold their parents liable for being inoculated?

If this kind of liability were to be accepted, where would that leave freedom of choice? And why do we tend to question parents' liability in one case (inoculation), but not in the other (braces)? Can we be sure that in both cases our train of thought departed from the parents' perfect freedom of choice, or do we view some situations in a more rational light than others?

The present contribution, however, does not aim to formulate an answer to all these questions, but rather aims to demonstrate how the new possibilities of human enhancement can cause a shift in fundamental principles that have been firmly established in our society for ages. It is generally accepted that parents cannot make a wrong decision regarding their child during and shortly after pregnancy. The idea that parents should be held liable for harmful decisions is quickly dismissed on the grounds that parents have a legal right to make their own choices based on their personal insights and beliefs. But how tenable is freedom of choice in a society where parents are increasingly required to make decisions concerning the enhancement of their child? And how can parents' freedom to choose according to their personal religion or belief hold its ground in a society where general consensus dictates the 'proper' course of action? If the legal right to freedom of choice remains intact, would it then nevertheless be possible for insurance companies to let parents face the financial consequences of their decisions?

Definition of Legal Concepts

In this section, I will first expound on a number of legal concepts before addressing questions of civil liability for parents' choices regarding the enhancement of their child. The function of tort law is to determine in which cases parties who have sustained damage are entitled to compensation. The guiding principle in determining liability is the unlawful act attributable to the tortfeasor—the person or organisation who broke civil law. An unlawful act can be firmly rooted in civil law statutes, but it can also result from unwritten law, namely the duty of care all individuals have towards each other. A tortfeasor is held responsible for a breach of a duty of care when he or she can be blamed for it. In addition, in some cases it is possible to attribute a fault to the tortfeasor even without proof of culpability, when there are compelling societal reasons for doing so. It is easier, for instance, to hold a big company or the government liable when they are not at fault: if they do not know the rules in detail, the consequences are at their own risk. Moreover, unlike the individual who has sustained damage, businesses and governments can pass on the costs of compensation to customers or taxpayers.

What is considered an attributable unlawful act depends, among other things, on place and time. In the late nineteenth and early twentieth centuries, railway operators could not be held liable if the sparks flying off a coal engine caused a farm to burn down. In addition to several other types of companies, the railway corporation was protected from liability because a high degree of liability would have been harmful for the development of a financially healthy company and for the construction of large infrastructures. However, towards the end of the twentieth century the emphasis shifted as it became more important to protect individuals against big companies—a trend that seems to be reversing again, tentatively moving towards less consumer protection and more freedom for companies.

Thus, what is considered an attributable unlawful act depends on the social context. The character of the unlawfulness is influenced by such factors as social relations and political and economic insights. The final piece of these considerations, however, is a legal judgement on the basis of which the concrete act is tested.

Since the breach of duty is at the core of liability law, it is impossible to determine liability based solely on the fact that damage has been caused. We daily suffer damage from human presence and behaviour. If baker John lures away baker Sam's customers by baking better-tasting bread, for example, baker Sam suffers damage. Nevertheless, in Western society it is evident that baker Sam is not entitled to compensation. Hence, the question of damage is never a leading principle in determining liability, nor does the severity of the damage have impact on the decision of whether a fault has been made. Instead, the main question at issue is whether the conduct that caused damage can be considered a breach of duty.

Imagine a woman who, while helping her sister move out of her apartment, gets her arm stuck between a cupboard and a wall after her sister trips and pushes the cupboard away from her the moment she falls. As a result of complex regional pain syndrome (CRPS), the woman's arm needs to be amputated. Clearly, the damage that has been caused is severe, but this does not make the sister's fall an unlawful act. In this case, therefore, the damage is to be paid by the woman who lost her arm or, if she has insurance, by her insurance company.

In order to answer the question of liability for a fault, it is important to know there is a clear difference between criminal law and civil law. Unlawful acts that fall under the domain of civil law include breaches of a duty of care, which usually cannot be considered as a crime or offence and are therefore not punished. An unlawful act in criminal law, however, often also comprises an unlawful act in private law. In other words, civil liability is more common than criminal liability.

Civil law exists to compensate for disturbances in the patrimonial balance between citizens (natural and legal persons). Criminal law intends to punish socially undesirable behaviour and to prevent it by the threat of punishment. In criminal cases, the state always acts as prosecutor on behalf of the general interest. The differences in set-up and goals between civil and criminal law are reflected in the sentencing guidelines. Criminal sentences such as fines are meant to punish the perpetrator by inflicting harm. Civil law, on the other hand, intends to restore the balance, not to punish or impose penance. Civil compensation for damages is therefore not a form of punishment, but rather a means to restore the victim as much as possible to the state that existed before the duty of care was violated.

The Duty to Inform Parents Regarding the Enhancement of Their Child

The dilemma between freedom of choice and liability in law is especially relevant in the case of parents. While health professionals and doctors also play an important role in the decision process, their duties and obligations can be determined quite precisely and their range of choices can be measured by objective standards. To a midwife, for instance, the question of whether or not to provide parents with information on the possibilities of prenatal diagnostics—after having been informed they have a family history of an inherited condition—is not a matter of personal beliefs but rather of duty. The parents, in turn, have to decide on whether or not to avail themselves of the available instruments. If the midwife does not notify the parents about their possibilities, he or she is at fault, and if the parents suffer damage as a result of their lack of knowledge they can hold the midwife liable for not offering them the possibility to choose.

This was true in the widely debated ruling of the Dutch Supreme Court in the case of a girl named Kelly. During the pregnancy, Kelly's parents had told the midwife about a family member with a severe congenital defect, but the midwife ignored this information and instead reassured the parents by saying that their previous child had been born perfectly healthy. When the child turned out to be handicapped, the Dutch Supreme Court decided that the midwife had acted unlawfully by not informing the parents about the possibilities of performing prenatal diagnostics. Moreover, the Dutch Supreme Court held the midwife liable for the damage that her unlawful act had caused the child. From this ruling it can be concluded that a child has the right to properly informed parents. This does not mean, however, that the child is entitled to a specific choice. On the contrary, the Dutch Supreme Court emphasises the immunity of the parents' choice:

> Nor can it be said that the judge, by invoking his authority [to calculate the damage in a manner that is most in accordance with the nature of the damage] (as described in article 6:97 of the Dutch Civil Code), brings closer or even creates the possibility for children in Kelly's position to hold their parents, or at least their mother, liable for their existence. After all, Kelly is not entitled to her non-existence, nor did she have a right to the termination of her mother's pregnancy. (Dutch Supreme Court, March 18, 2005. *Neder-landse Jurisprudentie* 2006, 606, para 4.16)

It is different when the choice concerns the performance of postnatal screening. In that case, children who suffer damage and think their parents made the wrong decision cannot claim a right to non-existence, but instead claim that, given their existence, they had the right to a better choice by their parents. Despite this difference, parents are nevertheless allowed to make a decision based on their beliefs of what is right, and their child cannot hold them legally liable. The midwife is obliged to offer parents the possibility to perform postnatal screening and the infant welfare centre is obliged to provide inoculations, but it is up to the parents to decide whether to accept or decline the offer.

In law, the notion of liability for the existence or non-existence of a child—handicapped or not—is undisputed in the case of doctors, midwives and other paramedics who did not carry out their duty to inform the parents. If it can be proven that they have not informed the parents when they should have done so, professionals can be held liable for any damage caused by their unlawful act.

The question is whether this also applies to the parents' liability. As stated previously, the liability of a paramedic or a hospital is dependent on unlawful acts that can be measured objectively, whereas the liability of the parents is related to the consequences that follow from a decision based on their worldview and beliefs. Hence, the question arises whether an essentially acceptable choice should be looked at differently in light of negative consequences. In principle, it is impossible to objectively assess the correctness of the choices described in this chapter. Had this been the case, we would no longer speak of choices but rather of duties which may or may not be enforceable by law.

Can a Free Choice be Wrong?

It is often assumed to be impossible to assess a choice made by parents on the basis of legal standards. In other words, a choice that parents are entitled to make can never be wrong. This also means that parents can never be liable towards their child. The right to choose is—within certain limitations—grounded in our fundamental rights. As a result, it is impossible to claim damages from parents on the basis of their decisions. This view gives precedence to the parents' freedom of choice over the child's right to hold them liable for these choices.

Another point of view, however, is that every choice comes with certain responsibilities. Everyone is allowed to make choices, but nobody is exempted from the liability that follows from these choices. When looked at this way, freedom of choice entails a consideration of the likelihood that a choice will turn out well or badly, and anyone who makes a decision automatically accepts the possibility of liability.

The dilemma is complicated by the fact that we are not dealing with clear-cut oppositions. After all, it is possible to make a choice and at the same time bear in mind that you can be held responsible for any possible harmful consequences. For example, if parents decide against terminating a pregnancy while knowing that their child has a severe congenital defect, they could take into account that their child might hold them liable and take out insurance against this risk. Thus, the possibility of being held liable does not fundamentally block the possibility of making a free choice, but it does take away part of that freedom. In the future, this might lead to parents being inclined to base their choice on what is considered acceptable by the majority and thus deemed to be 'right'. This, however, begs the question of how the prevailing opinion is to be determined. What should be our position on changing views and insights that cause us to view a case differently in retrospect than at the time of choosing? And how will such a development function in relation to the basic principle that parents should be free to make a decision in accordance with their personal views and beliefs?

Moreover, civil law requires the violation of a norm in order to assign liability; a fault must have been made. Therefore, parents can only be held liable after it has been ascertained that they were at fault. While technically it cannot be ruled out, however, the assumption that the parents were at fault is not compatible with the idea that they were free to make their own choice. After all, citizens are not free to act illegally. This becomes clear, for example, from the fact that a citizen about to commit an unlawful act can be prohibited from doing so.

If steps were to be taken towards the acceptance of children holding their parents liable for their choices, the field of human enhancement would influence liability law. Parents' choices regarding the existence and welfare of their child made on the basis of their personal views and beliefs would no longer be unassailable and would be treated in a similar way to other choices. Consequently, choices based on personal beliefs and worldviews would no longer be considered sacrosanct. This might in turn influence the nature of the choices made regarding human enhancement.

Influencing 'Free' Choice

It is interesting to explore in what ways the government can influence choices that concern the private sphere. For example, in 2007 the policy of the Dutch Government aimed to reduce the number of abortions by stimulating adoption. Can a child who was born with a congenital handicap hold the government liable because he or she would have never been born were it not for this policy? Is it possible to claim that such a policy violates the fundamental rights of children (e.g. personality rights or the right to being healthy)? Can a child who is adopted and unhappy hold the government liable for enforcing a policy that turned out to be harmful in his or her case?

Even more important is the question of how insurance companies can and will respond to the parents' choices. Imagine a woman who becomes pregnant at the age of 39. She has a higher risk of having a baby with a handicap such as Down's syndrome than a 16-year-old, but decides against prenatal screening. She knows that certain risks are associated with the performance of prenatal diagnostics, and she does not want to risk a miscarriage. After all, she has already decided to carry the child to term even if it does turn out to have a congenital disorder. The child she gives birth to is diagnosed with Down's syndrome and has to undergo a cardiac operation immediately after birth, resulting in an expensive bill that must be paid.

In order to answer the question of whether insurance companies should be allowed to attach consequences to such choices, the social context in which the choice is made should also be considered. It is not improbable that there will come a time when the chosen course of action of the 39-year-old woman will be considered unusual or even undesirable or offensive. From the point of view that available options must also be used, it might be argued that the woman is taking an irresponsible risk, and her choice to keep the baby regardless of the outcome may also be considered abnormal. Moreover, the majority may hold that the woman is burdening society with unnecessary costs. This would alter the principle that every choice concerning an unborn child, as long as it lies within legal boundaries, is equally correct. A change in social notions may gradually influence the consequences that follow from certain choices. In this context, an insurance company may also rethink their policy regarding the woman's choice, and for example decide to alter their insurance policy by adding a clause stating that they do not cover negative consequences that follow from a choice which is clearly considered abnormal by the majority. The woman would then still be allowed to make the same decision, but if her child turned out to be disabled, she would have to pay the costs herself.

Such a development can be blocked by the insurance companies themselves or by the legislature. In the long term, however, we must bear in mind that the legislature primarily takes into consideration the opinion of the majority, and might therefore not necessarily appreciate the value of minority opinions.

The Future of Free Choice

The dominant opinion on what is the 'right' choice under certain circumstances is influenced by the options available to gain insight into or influence a child's state of health, such as prenatal screening and preventive inoculation. The more options there are, the more we tend to think we should avail ourselves of them in order not to waste an opportunity, in particular where the medical sphere is concerned. Liability law deals with these tendencies by taking as its point of departure the parents' right to base their decisions regarding the welfare of their child on their personal beliefs, but this position may not hold forever.

The pressure facing the notion of free choice is reflected in law. In civil law, this pressure is formulated in the question of whether parents can be held liable for the damage they have caused their children by their medical decisions during or shortly after pregnancy. So far, liability law has answered this question with 'no'. This answer appears to be so self-evident that little attention has been paid to the ways in which this principle can be altered. Is it really that unthinkable to let parents pay the costs that come with a handicapped child if they consciously decided to have that child? Why should an insurance company have to pay these costs, and consequently divide them among a larger group of insurance payers?

Opposing this sanctification of free choice regarding medical proceedings during and shortly after pregnancy is an argument that is by no means uncommon in our culture: now that we have more knowledge, we have more options to choose from and can therefore also consciously choose to reduce certain risks. If we nevertheless decide to accept these risks, we should take responsibility if they become reality and pay for the costs ourselves.

The idea of deliberately taking a risk and expecting the insurance company to pay the costs in case that risk becomes reality is not, in fact, unusual. For example, the decision to drive a car intrinsically involves the risk of an accident. Thus, every driver accepts the risk inherent to the act of driving. If a driver is involved in an accident and sustains an injury, his or her insurer will pick up the bill—unless the accident was caused deliberately or through gross negligence. Health insurance companies do not argue that driving a car involves a consciously taken risk, that causing an accident can be considered an unlawful act, or that their clients should therefore pay their own medical expenses.

All things considered, it comes down to whether the risk is viewed as a collective or an individual concern. In the latter case, we are more inclined to agree that it is the individual who should pay for any damage their decision has caused, either to themselves or to others. However, between citizens damage is only compensated when caused by an unlawful act. Moreover, we cannot pronounce a value judgement on a decision regarding a child's existence and state of health that is based on personal beliefs. Choice A is not necessarily better than choice B, so neither can be wrong.

Only when parents' choices are no longer considered sacrosanct will it become possible for a child or an insurance company to hold them liable for their decisions. Whether this will ever happen depends on the value that society attaches to the freedom to make choices that deviate from the norm.

As mentioned previously, the perceived indisputability of the notion that parents are never liable for their choices regarding their unborn or newborn child is peculiar, because no indisputable arguments are given to support this view. While liability for not providing a child with braces does not meet with a lot of resistance, liability for not vaccinating a child does. My explanation for this phenomenon is that personal beliefs in general and religious beliefs in particular play an important role in the choices made during and immediately after pregnancy. In this respect, the question of whether or not to provide braces is free from value judgements, rendering it rather easy to judge that it is unlawful not to give a teenager braces when they are medically necessary. The opinion that it is wrong not to vaccinate a child may be thought in silence, but is not often expressed explicitly. After all, we are not allowed to pronounce a value judgement on decisions that are based on personal beliefs. Thus, our inclination to respect religious beliefs and the choices based upon them may, for the time being, guarantee parents freedom of choice and immunity from liability.

Even if it remains formally possible to make unusual choices (as choices made within the boundaries of the law cannot be considered unlawful nor lead to parental liability), the possibilities regarding human enhancement and the influence of social norms will always be important factors in the parents' decision-making process. This is normal and inevitable. When vaccination of children was introduced for the first time, parents had to make a conscious decision to inoculate. Over the past few decades, this situation has changed: gradually, the decision *against* vaccination became a conscious one. The unspoken norm was to inoculate a child according to the schedule offered by the government, and the decision against vaccination was associated with strong religious beliefs. Nowadays, this picture has become more diffuse. While inoculation still remains the norm, the reasons against vaccination have become more diverse, including not only religious beliefs, but also different views on the ideal development of a child's resistance or the risks of vaccination. Therefore, the individual choice of whether or not to inoculate a child is very different today compared to the situation in the 1940s. At the same time, however, it is considered important that parents should be allowed to make their own decision on whether or not to vaccinate without having to fear legal consequences.

Conclusion

I return to Paul, Evelyn and their son Timmy who was born in 2010. Paul and Evelyn decided to inoculate Timmy against bacterial meningitis one year later than usual. In the meantime, however, Timmy contracted meningitis, and it is clear that this will have consequences for his mental and physical development. Who should be presented with the bill?

11 Parents' Responsibility for Their Choices Regarding the Enhancement

Answer 1:
We assumed that Paul and Evelyn were free to make their own choices. Moreover, they were properly informed and made a calculated decision. They can obviously not be held liable towards their child.

Answer 2:
While Paul and Evelyn were free to make their own decision, this does not imply they are not responsible for any negative consequences that may follow from this decision. It is not unthinkable, therefore, that their child will hold them liable for the damage.

The most common view is that parents should carry the damage regarding any extra non-insured costs that result from Timmy's life, whereas Timmy himself carries any mental damage such as pain, sadness and laboriousness that may follow from being handicapped. Some costs will be compensated for by the insurance company. Under no circumstances can Timmy hold his parents liable, but how tenable is this?

The movements that can be distinguished in our current society—and which I have described above—do not all point in the same direction. The common notion that parents should not be liable for their choices during or shortly after pregnancy does not follow strictly from the principles of liability law. Great store is set by the principle of freedom of choice: choosing belongs to a person's private domain, which neither fellow citizens nor the government are allowed to enter. However, when a situation does not immediately concern ourselves, we can be very fast in judging that someone should carry both the negative and positive consequences of their choices. Moreover, in such cases, we are also inclined to think that while making an unusual choice should be allowed, any damage caused by that decision proves it to have been a wrong one. This can lead to an increased responsibility for the negative consequences of such choices. If a situation concerns ourselves, however, and if these negative consequences are presented to us rather than to someone else, our argumentation becomes the exact opposite and we do everything we can to make sure somebody else is held liable for the damage.

In light of these tensions, it is impossible to predict the answer to Timmy's case, but it is clear that the current situation cannot be considered a 'natural given'. Moreover, the different principles involved (free choice, a free choice is never wrong, responsibility for your own choice, damage should only be compensated in case of an unlawful act) can be explained in more than one way. Elaboration upon these principles, however, is anxiously avoided. I suspect that this is related to the fact that the choices mentioned in this chapter are often ingrained in religious or other strong personal beliefs: it is still taboo to pronounce an explicit judgement on such choices. By the time this changes, the approach to liability will also change, most significantly in the area of free choice—in the sense of 'free' from the risk of being considered an unlawful act. The most important factor in this development will be social pressure. Overall, it is by no means certain that the act of choosing will always be exempt from moral judgement, and as a consequence from far-reaching liability.

The increasing ways and means of human enhancement expand the range of choices, thus evoking the necessity to make certain decisions. In addition, the omnipresence of human enhancement renders it likely that one of the available choices will become the general standard, and it will take both time and awareness to protect all choices that deviate from this standard. The step from damage to unlawful act is smaller than we would like to believe, especially when choices based on religious or similar personal beliefs are no longer sheltered from external judgement. Parents who decide to postpone or abstain from vaccinating their child are often considered strange and peculiar. These parents only stand a chance if we all collectively decide—and that is also a choice—to protect them and their peculiarities. This is not a new problem, but one of all times and all cultures, as is the necessity to be consciously aware of this fact. We will be most successful at this if we keep in mind that tomorrow it could be us taking the plunge into the unknown depths of a non-standard choice.

Bibliography

Buijsen M (ed) (2006) Onrechtmatig leven? Valkhof Pers, Nijmegen
Kortmann SCJJ, Hamel BCJ (eds) (2004) Wrongful birth en wrongful life. Kluwer, Deventer
Sieburgh CH (2005) Schadevergoeding én leven. Compositie met rood, geel en blauw. WPNR 6637:755–762

Chapter 12
Concerning 'Humans' and 'Human' Rights. Human Enhancement from the Perspective of Fundamental Rights

Bert-Jaap Koops

Abstract This chapter investigates how, in the long term, human enhancement relates to fundamental rights. The right to equality and the right to vote, for example, can be applied to enhanced humans, although it will not always be easy to determine when a distinction is justified between enhanced and non-enhanced humans. New fundamental rights may have to be created, e.g. a right to identity, to mental integrity or to forget. In the long term, we need to determine whether robots and androids, if they function in ways comparable to natural or legal persons, could also claim legal protection through fundamental rights. Fundamental rights should also steer the development of human enhancement. Individuals have a right to improve themselves, but they must be able to resist enhancement as well. We could consider introducing a fundamental right to imperfection, to ageing and even a right to die, as well as extending the government's duty of care to promote human diversity. Since enhancement can be a right but never an obligation, fundamental rights will have to play an important role in preventing 'normal' humans from becoming an underclass to enhanced humans.

London, 28 June 2079, from our reporter.
The large demonstration of transhumans that took place on Trafalgar Square yesterday proceeded peacefully considering the circumstances. Around 800,000 robots and androids answered the call of the Transhuman League to demonstrate for the recognition of fundamental rights for their kind. 'We want to have our rights recognised after all this time.

Translated by Lotte Oostebrink and Lydia ten Brummelhuis.

B.-J. Koops (✉)
Tilburg Institute for Law, Technology and Science, Tilburg University, P.O. Box 90153
5000 LE, Tilburg, The Netherlands
e-mail: E.J.Koops@uvt.nl
http://www.tilburguniversity.edu/webwijs/show/?uid=e.j.koops

B.-J. Koops et al. (eds.), *Engineering the Human*, DOI: 10.1007/978-3-642-35096-2_12,
© Springer-Verlag Berlin Heidelberg 2013

We too have the right to live', says Andy02593, a third-generation android. 'My built-in on/off switch is very humiliating; it restricts my freedom to develop myself'.

The jubilant mood and atmosphere of solidarity were somewhat tempered by a considerable counterdemonstration of humans, led by the Appeal to Human Dignity. The front human of the AHD, Christian Flatfoot, spoke for many in his speech: 'Transhumans are different from humans. They are very useful for mankind and the world, but that does not mean they should simply be given all kinds of rights. Imagine androids receiving passive suffrage and ruling the country. Before you know it they will merge United Europe with the Asian Union and humans will slowly be phased out. It is absolutely vital that transhumans remain subordinate to us, for the protection of mankind'.

Although the AHD has a firm footing in society, it is expected that the increasing demand to provide rights for transhumans will be granted by the government. The Minister of Justice Warwick, cloneson of the colourful UK scientist of the early twenty-first century, is rumoured to be preparing a bill to include rights for transhumans in the Constitution.

This Web message may appear absurd to many readers today. The theme of cyborgs and robots as a new class in society can be found in science fiction, but it is more associated with fiction than with science. But the idea is not that absurd. If, 300 years ago, someone had written an essay that started with a newspaper article from 1948 about a demonstration of slaves, women and children claiming human rights, many would have called them mad. To us it seems self-evident that black people, women and children are as much human as white men, and therefore have human rights. Why should it be impossible that in the future cyborgs and androids will also be seen as humans?

There is much to say about this, one of the most fundamental questions arising from the development of human enhancement. Human enhancement has many manifestations, from IVF babies to cyborgs and beyond, all of which evoke legal and regulatory questions. In this chapter, I will investigate the significant forms of human enhancement that may play a role in the longer-term future from the perspective of fundamental rights. My central question is: how will the increased enhancement of humans and fundamental rights relate to each other in the long term?

To expand on this central question we can ask two further types of questions. First, how can or should the existing fundamental rights be applied to enhanced humans, and are new fundamental rights required to protect enhanced or non-enhanced humans? And second, if the application or applicability proves to be problematic, should fundamental rights then be adapted to enhanced humans and/ or should human enhancement be adjusted according to fundamental rights?

These two types of questions may be answered in a more or less consecutive manner. As enhanced humans evolve and the application of fundamental rights is increasingly difficult, the demand to adapt fundamental rights and/or human enhancement becomes more pressing. Moreover, there is also a certain chronology implied in the second type of question. At first, fundamental rights will operate mainly as a guiding principle with regard to enhanced humans, but at a certain point—as enhanced humans continue to evolve—fundamental rights will move with the times and be adapted themselves. When this transition point will occur depends on many factors—not only on technological possibilities but also on ideas about fundamental rights (whether they are visionary and forward-looking, or moderate and consolidating) and on ethical, cultural and religious views.

Since my intention is not to predict or design a constitution for 2079 but to reflect on the implications of human enhancement for fundamental rights in the coming decades, I will put a relatively strong emphasis on the first type of question, the applicability of fundamental rights to enhanced humans, and pay slightly less attention to the implications of human enhancement for fundamental rights. However, I will briefly discuss this last type of question to show that fundamental rights and the underlying values they reflect are not fixed but defined in interaction with developments in society.

After providing short background sketches of human enhancement and fundamental rights, I will investigate three aspects of the central question, starting with an overview of the existing fundamental rights and how these relate to human enhancement. Next, the question arises whether human enhancement requires the development of new fundamental rights, such as a right to ageing. Finally, I will elaborate on the fundamental question already mentioned: in the long run, who will have fundamental rights, also known as human rights?

Human Enhancement

The term 'human enhancement' is used to denote a variety of processes or outcomes thereof. These roughly have in common that they aim to permanently change or enhance humans—individuals or 'mankind'—through non-'natural' additions to or alterations of their essential characteristics. Some remarks can be made about terms such as 'enhance' and 'natural' in this definition, but I will not go into this here: Christoph H. Lüthy covers it in his contribution to the present volume (Chap. 2). I will use 'human enhancement' as an umbrella term for this phenomenon and employ the term 'enhanced humans' to define the resulting products.

Human enhancement includes processes that range from plastic surgery and tissue engineering via the selection of wanted or unwanted genetic characteristics, genetic engineering and the implementation of neural implants, to human–machine systems and cyborgs. I will mainly look at 'enhanced humans' in the somewhat remote future, meaning those kinds of creatures whose 'human' character is called into question. Clearly, these types of beings are the most interesting in the context of human rights. In this contribution the following, overlapping, characters will be dealt with:

- Human: Species of animal that calls itself 'human' and has not been able to develop a commonly accepted definition of this term;
- Quasi-human: A humanlike individual that does not fit within the definition of human (as applied in the relevant context);
- Enhanced human: A human or quasi-human that is the product of human enhancement, including:
  - Android: A robot with built-in human characteristics
  - Chimera: A combination of a human being and animal(s)
  - Cyborg: A half-human, half-machine hybrid, created from a human with increasing amounts of added technology
  - Clone: A human created through reproductive cloning
- Entities with legal personhood:
  - Natural person: The legal term for a human individual, to be distinguished from a legal person
  - Legal person: A legal denotation of an entity which is not a natural person but has the legal capacity to act.

Fundamental Rights

Human rights are often called fundamental rights, because they are the most basic and fundamental rights available to human beings. Since legal persons can also claim human rights and because it is somewhat contrived to expand the definition of 'human' to include legal persons, it is more appropriate to use 'fundamental' rather than 'human' rights. These rights are necessary to protect citizens in a democratic society governed by the rule of law—originally against the government, but in the last decades increasingly against fellow citizens or companies as well. Moreover, these rights are an essential tool to develop oneself in society.

Fundamental rights are codified in the form of constitutions and various international conventions. Important international instruments are the European Convention on Human Rights and Fundamental Freedoms, the Charter of Fundamental Rights of the European Union ('European Charter'), the International Covenant on Civil and Political Rights and the International Covenant on Economic, Social and Cultural Rights. The enumerations and formulations of fundamental rights in these conventions and in national constitutions have many similarities, but there are differences as well. As the aim of this essay is not to give a comprehensive analysis of all fundamental rights but to explore their implications, I have opted to restrict myself in this essay to the Dutch Constitution, which conveniently contains a representative overview of fundamental rights from which I can draw relevant examples.

Human Enhancement and the Existing Fundamental Rights

What questions can we expect in the coming decades regarding human enhancement as seen in the light of the existing fundamental rights? I will briefly discuss Chap. 1 of the Dutch Constitution and highlight several fundamental rights as examples. The first article is immediately relevant:

> All persons in the Netherlands shall be treated equally in equal circumstances. Discrimination on the grounds of religion, belief, political opinion, race or sex or on any other grounds whatsoever shall not be permitted.

There is nothing wrong with differentiation; we encounter unequal cases that call for different treatment on a daily basis. However, when that differentiation is unjustified it becomes discrimination. The crucial point here is what is included in the term 'all persons'. It refers to all living resident humans in the Netherlands, but not to animals or legal persons. As long as enhanced humans are regarded as human beings and live in the Netherlands, they are able to claim this right to non-discrimination. I will return later to the question of when an enhanced human should be regarded as 'human', but it is clear that Article 1 is fundamental in establishing how our future society will handle enhanced humans. As soon as a type of enhanced humans is not qualified as human, a dichotomy between humans and quasi-humans will arise.

Discrimination against quasi-humans will then be allowed. This does not have to be a bad thing. For example, we currently have a dichotomy between humans and animals, and there is hardly any discussion whether it is morally correct for animals to be discriminated against. But when quasi-humans increasingly resemble humans without qualifying for that definition, the dichotomy will cause tension and in time might lead to demonstrations on Trafalgar Square. This issue I will return to later. For the remainder of this section, I will discuss enhanced humans who are 'human' and thus entitled to fundamental rights in order to investigate what these rights mean for them.

In the short term, it is relevant to ask whether discrimination on the basis of enhancement is allowed. This concerns enhanced humans who do meet the definition of 'human', but still are different. This could be, for example, because their ancestor's gametes were genetically manipulated, because they are clones or because they have a bionic arm, a brain implant with an Internet connection or a longer life expectancy. Since the prohibition of discrimination involves 'any other grounds whatsoever', enhancement will have to be included. In my opinion, this will be especially important for the medical and insurance industries, because there may be substantial differences between the assessment of the health risks of humans and enhanced humans. Nevertheless, it is still possible to make a differentiation in those areas. It is not by definition unreasonable to demand higher liability insurance premiums from someone with a bionic arm, if that person runs a higher risk of causing damage. Thus, in some respects enhanced humans can be treated differently, but as long as this is done on the basis of a relevant difference, we need not fear discrimination against them.

However, I think the opposite kind of differentiation is more probable and potentially more dangerous. When enhanced humans function better, are more beautiful and more intelligent—which is, after all, often the intended outcome—non-enhanced humans will quickly be left behind. The right to equality will play a vital role in protecting 'normal' humans from becoming an underclass to enhanced humans. This does not go without saying. Again, differentiations can be justified. If an employer can choose between a human with an IQ of 130 and an enhanced human with an IQ of 210, why should she not be allowed to choose the more suitable candidate? In other words, is it legitimate for her to make a distinction on the basis of a quality that is the consequence of the *non*-enhancement of a human?

Precisely because this question is difficult to answer, the prohibition of discrimination will have to serve as a guiding principle in the development of human enhancement. If we want to prevent humans who choose—if this choice will even exist—not to participate in the enhancement trend from becoming an underclass, society will have to set limits to the ways in which humans can enhance themselves. If brain improvements lead to a structural distinction between 'ultra smarts' and 'simple minds', this could well be a reason to ban brain improvement until it is accessible enough for everyone to benefit. On the other hand, brain improvement can also provide great benefits to society, and if small-scale initiatives are banned, it may never get off the ground. Therefore, politicians must weigh the options between economic progress and self-development on the one hand, and equality and solidarity on the other.

Besides, we have to realise that differentiation is not always the result of a conscious decision. Young, attractive Caucasian men can have an easier time finding a job than old, unattractive immigrants, even when the qualifications of the former are objectively lower. Psychological and cultural factors play an undeniable role in decisions that involve humans, even though this role is often invisible and subconscious. A legal right to equality has little effect in these instances. Especially when enhancement is used to boost favoured cultural characteristics—symmetry, tallness, fatness, androgyny, whatever is considered attractive in that place and time—other mechanisms apart from legal ones will have to be developed to guarantee that non-enhanced humans are treated fairly.

The next two fundamental rights evoke similar questions: the right to be 'equally eligible for appointment to public service', meaning the right to become a civil servant (Article 3 of the Dutch Constitution), and active and passive suffrage (Article 4). These rights apply to 'all Dutch nationals', so if individuals want to claim them, they do not only have to be human, but also Dutch. I foresee few problems here. Indeed, it is to be expected that enhanced humans who are regarded as humans by society will fall under the Dutch nationality act.

However, questions similar to those concerning the right to equality do arise regarding enhanced humans who are, although still human, considerably different—and better—than non-enhanced humans: can the latter be 'equally eligible' if enhanced humans will structurally be preferred as civil servants? What is the point of passive suffrage when enhanced politicians, with their improved beauty, intelligence and genetically generated charisma, have a structural advantage over non-enhanced politicians? On the other hand, is there really such a great difference between enhanced humans and current enhanced politicians, who are surrounded by PR professionals, media trainers and plastic surgeons? This depends on the degree of enhancement and more importantly on the individual's freedom of choice to go along with the enhancement into *homo politicus perfectus*. Equal eligibility for appointment ends when a class of ideal politicians is bred in test tubes.

One of the most important fundamental rights in relation to human enhancement is the right to inviolability of the body (Article 11 of the Dutch Constitution):

> Everyone shall have the right to inviolability of his person,[1] without prejudice to restrictions laid down by or pursuant to Act of Parliament.

This right was added to the Constitution in 1983 to particularise the right to privacy (Article 10) and aims to repel actions that violate physical integrity. This particularly concerns actions of governmental bodies, such as taking buccal swabs for DNA profiling, strip or body cavity searches or shaving off a beard for an identity parade. With regard to human enhancement, it means that the government cannot simply enforce the enhancement of humans, for example by making the use of chip implants or mood-changing pills mandatory.

'Not simply' indicates that the government would, in fact, be able to do this by passing a specific law. Such laws may allow all kinds of interferences with physical integrity. For example, if biotechnology develops in such a way that the ageing process of cells can be delayed or even stopped via a small operation, the government could make this operation mandatory in order to increase the age of retirement and thus productivity. However, this obligation must conform to Article 8 of the European Convention on Human Rights: a breach of privacy is only allowed when it is regulated by law, necessary in a democratic society, and 'in the interest of national security, public safety or the economic well-being of the country, for the prevention of disorder or crime, for the protection of health or morals, or for the protection of the rights and freedoms of others.'

Staying young will benefit both health and the economy, but whether it is necessary for society to make it obligatory remains to be seen. In the end, this is a political decision that must be tested before the European Court of Human Rights. The international context will play a role as well. Should other parts of the world rapidly develop certain types of human enhancement, causing economic devastation in Europe, then enhancement legislation would be easier to justify. However, in such a scenario discussions about the post-human economic world order in relation to human rights would already have taken place within the UN, WTO and

---

[1] The official translation reads 'person', but that suggests a wider scope than Article 11 actually protects, namely only the physical integrity of the body.

other international forums—discussions which the EU could exert its influence on. For the moment I think it improbable for government to make enhancement in whatever shape or form mandatory, although we cannot rule out this possibility when one day a Transhumanist Party should be in government.

Apart from a negative right to be safeguarded against violations of bodily integrity by others, Article 11 of the Constitution also guarantees the positive right to self-determination: the right to determine what is done to one's body. This is interesting, also in the short term, because in a sense it suggests a right to enhancement—at least for the body. Anyone wishing to pierce their skin with jewellery should be able to do so; a ban on piercings would probably not survive scrutiny by the European Court. This means that citizens in principle have the right to improve their bodies, in ways ranging from plastic surgery via bionic arms to brain implants. However, having the legal right does not mean that it should also be technically possible, cheap and without risks: the right to physical integrity does not imply a positive obligation for the government to try and make citizens' bodies as good as possible. But anybody who wants to be a cyborg, like Kevin Warwick (author of *I, Cyborg*, see Chap. 7, this volume), *is* allowed to do so. At least, until the government decides that there are limits to enhancement and that certain forms must be prohibited.

At the moment, reproductive cloning is forbidden (Article 3(2) (d) of the European Charter; cf. Koops, Chap. 10, this volume) and it is not unlikely that due to technical advances other forms of enhancement will also be prohibited if they are considered unethical or otherwise undesirable. Such prohibitions are probably permissible because they are in the interest of 'the protection of health or morals'. Especially when technologies interfere, or seem to interfere, with the essential characteristics of humans, the European Court will allow a prohibition on enhancement, and in those cases it could well be a long time—because of the wide margin of appreciation the Court grants national states—before the individual right to self-determination outweighs the national interest in the ban on a certain type of enhancement.

It is interesting to look at several specific questions concerning the fundamental right to physical integrity and the development of cyborgs. When, in the not too distant future, specific brain signals will be used to perform functions outside of the body, for example allowing a paralysed person to control a prosthesis or a cursor on a screen, the question will arise whether the police should be allowed to intercept such chip-controlled brain signals. This question becomes even more relevant if in the long term Warwick's vision of a future in which cyborgs and cyborg–humans communicate primarily through thoughts rather than speech should become reality. In this scenario, the interception of communication does not only breach the right to secrecy of communication (Article 13 of the Dutch Constitution), but also—and more importantly—the right to physical integrity. And should the police be allowed to perform a computer network search (art. 19 para. 2 of the Convention on Cybercrime) into the brain of a cyborg who is inextricably connected to the Internet?

A final question concerns the end of fundamental rights: death. Only living humans are entitled to fundamental rights, and the right to self-determination in Article 11 expires upon death. However, there are companies that offer to deep-freeze humans immediately after death and to call them back to life when this will become technically possible. At a company named Alcor, for example, one can have oneself 'cryopreserved' for a mere 150,000 dollars. The company describes this as 'the science of using ultra-cold temperature to preserve human life with the intent of restoring good health when technology becomes available to do so' (www.alcor.org). As of July 2012, Alcor had already frozen 112 'cryopatients'.

It is interesting to consider the legal-theoretical issue of the constitutional status of frozen bodies or bodily materials during the period between first and second life. According to the existing law, they no longer have a right to self-determination and can therefore be destroyed, for example in the case of a cryopreservation company's bankruptcy, without a breach of fundamental rights. The bodies can also be modified without consent—after all, the persons associated with the bodies are dead. This would mean that after defrosting and revival those persons would have undergone a modification they did not approve, which may radically impair their right to self-determination in second life. Since all of this is technically very hypothetical, we do not at present have to be concerned about the fundamental rights of frozen bodies, but this example does clearly show that human enhancement raises interesting questions concerning fundamental rights theory.

Finally, let us turn our gaze towards the social-economic fundamental rights in Chap. 1 of the Dutch Constitution: work and employment, social security, environment, health, housing, cultural development and education (Articles 19–23). The government has the responsibility to take care of these collective goods. Although this concerns duties of care, which do not require concrete results, it is clear that what the government can do or allow to happen is bound by the demands and limits set in these social fundamental rights. For example, if enhanced humans work so efficiently that the majority of the population become unemployed, the government has to intervene. Human implants must not have too great an impact on the environment. And the government might be able to use certain forms of enhancement to further the people's social and cultural development, for example by stimulating research into altruistic genes and genes that are associated with sensitivity to art.

To me, however, the most important article seems to be Article 22 Section 1: 'The authorities shall take steps to promote the health of the population.' Here the blurry distinction between therapy and enhancement becomes important. The government must stimulate public health—the healing, or making 'normal', of 'sick' people—but not enhancement—the improvement of 'normal' people by enhancing their 'normal' positive qualities. Many forms of enhancement fall under this second category, but some types of enhancement *are* aimed at public health. A case can be made that delaying or stopping the ageing process falls under therapy; while being old is not a disease, it is often accompanied by disease and defects. Staying young thus usually means the absence of illness (whether this is technically true for genetic anti-ageing interventions I will not discuss here). In this light, Article 22 could be read as a call to promote these kinds of enhancement, especially through the stimulation of scientific research.

Human Enhancement and New Fundamental Rights

New fundamental rights are not quick to come into existence, but the result of profound and extensive social, cultural and political developments. The right to data protection (Article 10 Sections 2 and 3) is an example of a new fundamental right, added to the Dutch Constitution in 1983 after an extensive debate and maturation process in several international legal instruments, influenced by the rise of computers and the increase in automated data processing.

It is useful to start thinking now about possible new fundamental rights that may be considered desirable or necessary due to the rise of human enhancement. The existing fundamental rights, as described in the previous section, give rise to a number of problems and questions when applied to enhanced humans, but are overall future-proof. However, this does not mean that they are also sufficient for a society in which human enhancement flourishes. On several points there may be a need for new forms of legal protection.

First, we should consider the rights of the enhanced humans themselves. Although they have an advantage over non-enhanced humans—if things go well, enhancement should make them less rather than more socially vulnerable—it is conceivable that new needs for legal protection will arise. For example, clones may be treated by society as a 'copy' of their 'original' and thus feel constrained in their development (see Koops, Chap. 10, this volume). A fundamental right to identity, meaning the right to develop oneself and to safeguard one's self-image from unnecessary outside interference, could repel such effects. I do not mean, in the words of Paul Ricoeur, a right to *idem*-identity, to sameness—this, as a right to name and nationality, can for example already be found in Article 8 of the Convention on the Rights of the Child—but a right to *ipse*-identity: to selfhood, which is important for self-development.

In the long term it is possible that androids, especially if they want to appear more humanlike in order for humans to treat them more seriously (see Chap. 7, this volume), will have a need to be able to laugh, cry or experience pain, and a fundamental right to emotions would fulfil this need. Such a fundamental right may also be relevant for humans whose capacity for negative feelings has been genetically removed (which might have the side effect of levelling off positive feelings as well). After all, negative feelings also contribute to the wealth of human experience.

In the short term, a right to mental integrity may be useful if the brain of an enhanced human will be connected in all kinds of ways, without a physical component, to the outside world. In Dutch law, the fundamental right to physical integrity historically only applies to actions that physically affect the body: in contrast to Article 3 of the European Charter, which covers both physical and mental integrity, mental integrity is only covered by Article 11 if it is affected by a physical action. The thought behind this is that the right to privacy (Article 10) also protects the mind; therefore, no separate protection is needed. When wireless brain communication will be introduced, however, this may change. The right to privacy may offer insufficient explicit protection to cyborgs who do not want to be intruded upon in their thoughts.

Furthermore, we must also consider a right to forget and be forgotten. This is currently not an issue for humans (although it may be for the Internet), because it is human to forget and everyone is affected by it. Indeed, it is useful to be able to forget and be forgotten. In a society where brain functions are expanded—aided by external storage units connected to the brain—to improve memory, it will become a lot harder to forget. This may for instance have undesirable consequences when coping with traumatic experiences. A right to forget could then become relevant.

However, a right to be forgotten seems more urgently needed. Society records ever more information in a variety of files and connected networks. Their whole lives, children of the 'digital natives' generation have left digital trails on the Internet: on Web pages and blogs, in news groups, chat rooms and online social networks. For the rest of their lives they may be confronted with something they have once, as a youthful sin, done, shown or said. If enhanced humans should possess direct and continuous access to the Internet—for example by using augmented reality glasses with face-recognition software that look up a profile on the Internet and display it (name, age, occupation) in the margin of their vision, or by means of a hand-held Internet computer connected to the brain—they would be able to trace the digital trail of anyone they meet and use it in their interaction with that person. The unrelenting digital memory thus connects a person to what they once were or said, even though they may have changed radically in the meantime. The prohibition of discrimination is not equipped to protect against the digital elephantine memory. The risk here is not so much an unjustified distinction in society but the freezing of someone's identity in their past. A right to be forgotten—whether or not in the shape of a ban for cyborgs to access other persons' life history unasked—could limit this risk.

Second, and in my opinion more relevantly, we should consider the rights of non-enhanced humans. In principle, in its guise as a right to self-determination, the right to physical integrity makes it possible for people not to jump on the enhancement bandwagon. Whoever is so inclined should in principle be able to remain a non-enhanced human. However, social reality will often be different. Certain types of enhancement will have so many benefits—including an interesting job, improved health, appealing leisure activities and, as a consequence, economic and social advancement—that people will indirectly be forced to take part in enhancement. When a majority of the population reaches a high age while remaining youthful through anti-ageing methods, the age of retirement will sooner or later go up and everyone will have to continue to work longer, including those non-enhanced individuals who have aged faster.

We could also question whether a voluntarily chosen longer and healthier life leads to that much more happiness. When you have a life expectancy of 165 and a retirement age of 120, at 90 you may well be tired of starting yet another job after yet another outsourcing project. The prospect of having to go on for another 30 years may have little appeal. To prevent society from dragging everyone into a semi-compulsory rejuvenation cure just for the economic and social benefits of a longer and more youthful life, a right to ageing might be considered. Anyone who wants to stay young may choose to do so, but those who want to grow old and cut back on life should also be given that possibility.

Another consideration, as an extension or alternative, could be a right to imperfection. Human enhancement will potentially evolve in the direction of a perfect human image, and when enhancement is both cheap and accessible there will be a high social and possibly economic pressure to smooth away slight imperfections. It is standard policy today to adjust crooked teeth, but a few decades ago it was common practice not to treat crooked but otherwise healthy teeth. The enhancement of the future will be more far-reaching. Embryo selection, genetic manipulation and other forms of intervention in the essence of human beings will eliminate 'undesirable' qualities. Nobody objects to this in the case of Huntington's disease or other serious afflictions, but what should happen in cases such as genetically induced deafness, albinism, colour-blindness, obesity, ADHD, left-handedness, red-headedness or homosexuality? This spectrum of personal characteristics, commonly accepted as ranging from disease to normality, shows that the distinguishing line between disease, disorder, deficiency and normality is fuzzy. When we start selecting against one end of the spectrum (Huntington's), this may in the long term lead to selection against normal, but socially awkward, qualities until we are left with a 'perfect', uniform population. The slide down this slippery slope could be stopped by a right to imperfection or to being different. Individuals would then be given the right to remain deaf or colour-blind or ugly, to not replace an amputated arm by a realistic-looking bionic arm, and to other forms of what the majority would regard as imperfection. Others have the mirror obligation to respect this being-different, something that must be imposed in practice– easier said than done—through enforcing the law. That way, individuals would be better able to resist the pressure for perfection.

Following old age and disease is the even more important question of death. What if it becomes technically possible to postpone death ever more (see Chap. 8, this volume): does this mean that an essential characteristic of humans—mortality—becomes impaired? Robbert Dijkgraaf, former president of the Royal Netherlands Academy of Arts and Sciences, calculated that if humans were to become immortal, life expectancy would be around 2,000 years, at least in current society. Human beings may no longer die a natural death, but sooner or later an accident or crime will still end their lives. The prospect of becoming 2,000 years old—more than twice the age of Methuselah—may appeal to some, but also evokes an image of endless boredom and, as in the film *Groundhog Day*, not knowing what to do when you have seen and done everything over and over again. I think that the right to die, for example in the shape of a fundamental right to euthanasia or suicide,

12 Concerning 'Humans' and 'Human' Rights 177

will become an important fundamental right when enhanced humans live ever longer. However, such a fundamental right is currently not in existence, as the European Court of Human Rights decided in the case of Pretty v. United Kingdom in 2002.

Third, we could think of more collective fundamental rights that would benefit society as a whole. The social-economic fundamental rights may need expansion. The right to health provides a starting point to stimulate human enhancement, but more may be required. While some plead caution to prevent humanity plunging thoughtlessly into an unknown and potentially un-human future, others argue for a right to progression. They do not emphasise precaution, but proactivity:

> If the precautionary principle had been widely applied in the past, technological and cultural progress would have ground to a halt. Human suffering would have persisted without relief, and life would have remained poor, nasty, brutish, and short [...]. Most activities involving technology will have undesired effects as well as desirable ones. Whereas the precautionary principle is often used to take an absolutist stand against an activity, the Proactionary Principle allows for handling mixed effects through compensation and remediation instead of prohibition. The Proactionary Principle recognizes that nature is not always kind, that improving our world is both natural and essential for humanity, and that stagnation is not a realistic or worthy option. The Proactionary Principle stands for the proactive pursuit of progress. (www.maxmore.com/proactionary.htm)

In other words, some argue for a right to technological advancement or innovation in which experimentation is both allowed and stimulated and the risks taken are proportional to both the potential negative and positive consequences. A social-economic fundamental right to innovation and biotechnological progress is not at issue in the current political order. However, it is important to consider such a right in the debate, because human enhancement has undeniable advantages for society, which may not reach their full potential when hindered by too much precaution.

Stimulating the good does not mean that the bad must be banned. Alongside or instead of an individual right to imperfection, it may be necessary to establish a positive obligation on governments to stimulate the pluralism of humankind. In the same way that freedom of speech fosters the pluralism of information to ensure that social debate is fuelled by different points of view, so society also needs pluralism of people. To put it differently: we should consider a social-economic fundamental right to variation or imperfection. One of the reasons why American political scientist and philosopher Francis Fukuyama is concerned about the bio-technological revolution is the effect of perfection on being human. When enhancement has erased all negative qualities of humans, no positive quality will remain:

> [...] what we consider to be the highest and most admirable human qualities, both in ourselves and in others, are often related to the way that we react to, confront, overcome, and frequently succumb to pain, suffering, and death. In the absence of these human evils there would be no sympathy, compassion, courage, heroism, solidarity, or strength of character. A person who has not confronted suffering or death has no depth (Fukuyama 2002, pp. 172–173).

Pluralism of human qualities and the cherishing of imperfections may become important duties of care in the future, to preserve an essential element of our human existence—the depth and richness of human experience.

Legal Persons and Animals as Holders of Fundamental Rights

Apart from the existing and new fundamental rights, I would briefly like to investigate the third aspect of my question: who will in the long term be entitled to fundamental rights? Humans are not the only ones to have human rights; legal persons can also, to a certain extent, claim these rights. A legal person is a legal construction that makes it possible for an entity to perform legal acts. This may be a municipality, private company, church congregation or association. Such entities are usually represented in society by natural persons (humans), but these natural persons are not always personally liable for the actions they perform on behalf of the legal person. This legal construction simplifies social intercourse and governance.

Hence, legal persons have fundamental rights to a certain extent. When the Dutch Constitution was drawn up in 1983, the constitutional legislator declared that the fundamental rights guaranteed in Chap. 1 also apply to legal persons, and similarly to groups and organisations which are not legal persons, in so far as this is relevant given the nature of the fundamental right. For example, there is no point in talking about the physical integrity of a municipality, but a company facing legal prosecution does have a right to a fair trial. The scope of the legal protection is usually somewhat more limited than in the case of natural persons. A business or foundation may appeal to the right to inviolability of the home to safeguard against unreasonable searches of their premises, but since a business or foundation building is less privacy-sensitive than the home of natural persons, an infringement of the right to inviolability of the home is easier to justify in these cases.

Although there have been some calls to give animals fundamental rights in the Dutch Constitution, up until now non-human animals do not have fundamental rights. Animals do have legal protection (for example, through the American Endangered Species Act), but only as objects and not as legal entities. For example, when their rights have been violated, animals cannot go to court on their own, nor can humans or organisations do so on the animals' behalf. They are only able to request protection of animals through a judge on behalf of themselves.

When deciding what fundamental rights mean for the development of human enhancement, it is interesting to keep in mind the fact that legal persons can (to some extent), and animals cannot (at present), claim fundamental rights. In some cases, it is evidently useful for society to give fundamental rights to non-human entities. However, there does appear to be a threshold for providing fundamental rights to entities that strongly resemble humans in certain respects. This may be caused by the way these entities function in society. Animals are abundantly present in society, but they are—unlike legal persons—not expected to take part in social intercourse.

Enhanced Humans and 'Human' Rights

How will human enhancement affect the application of fundamental rights? Can we continue to use the term 'human rights' as synonymous to fundamental rights, or will this term gradually disappear as we need to place its first element between increasingly large quotation marks: 'human' rights? In order to answer this question, I will for the moment disregard the question of what constitutes a human being. Numerous distinguishing characteristics have been proposed to define humanity, including intelligence, self-consciousness, going on holidays, the ability to self-reflect and baking pizzas, and presumably there are few creatures in the universe that share this exact combination of characteristics with humans.

However, the defining essence of human beings is less relevant for the discussion about fundamental rights than is their function in society: to protect citizens from abuse of power and to safeguard their lives and development opportunities. This function is also important—although to a lesser degree—for non-natural persons who participate independently in society, implying that legal persons can claim fundamental rights as well.

In this functional approach, it seems natural to reply in the affirmative to the question whether enhanced humans can also have fundamental rights—that is, in so far as they are independent participants in society and need protection from abuse of power and to promote self-development. In this respect, there will be no fundamental difference between non-enhanced humans and enhanced humans that derive from *homo sapiens*, such as genetically modified humans, clones, cyborgs and human-based chimeras. While they may be engineered somewhat differently, in general they will perform the same kinds of functions and types of actions in society. It is well conceivable that fundamental rights, along with the image of 'humanity', will gradually co-evolve with enhanced humans and that in a 100 years' time it will in retrospect have been a curious question whether enhanced humans should be entitled to fundamental rights. Of course they should, for they are humans (in the twenty-second-century idea of 'humanity')!

With regard to enhanced humans that do *not* derive from *homo sapiens*, however, such as androids and other robots, the question is more difficult to answer. I expect that somewhere in the future—but not for several decades—there will be a turning point after which fundamental rights can no longer be withheld from robots and androids. After all, they will become increasingly active and independent participants of society—a trend that is gradually becoming visible in, for example, window-cleaning, traffic-conducting and vehicle-driving robots—and will perform more and more tasks and functions of humans. Sooner or later it will be desirable to acknowledge the android as an entity functioning in society in such a way that it needs legal protection. This does not immediately have to be at the level of fundamental rights, but certain fundamental rights will have an impact on the legal protection of androids. As in the case of legal persons, it will gradually become accepted that androids and certain types of independently operating robots can also lay claim to fundamental rights, in so far as this makes sense depending on the nature of the fundamental right in question. Exactly *how* and *which* fundamental rights will be applied will of course depend on the way in which the androids function. A crazed android will presumably be reprogrammed earlier than a crazed human will be brainwashed, but an intelligent and charismatic android will in principle be equally eligible for appointment to public service as any 'normal' human being.

In short, fundamental rights will in the future also be given to enhanced humans who derive from human beings and—in the longer term and to a certain degree— to androids and robots that function in society in a way comparable to natural or legal persons. The content and scope of these fundamental rights may differ according to the type of enhanced or non-enhanced human, but their claim to fundamental rights as such will not be disputed.

Conclusion

This long-term exploration of fundamental rights in light of human enhancement indicates that, as human enhancement continues to develop, there is a lot to discuss and decide. Although this still seems like a distant reality, it will inevitably come closer.

First of all, the development of human enhancement has consequences for fundamental rights. These will have to be applied to enhanced humans—that is, those types of enhanced humans that derive from human beings, such as genetically modified humans, clones, cyborgs and certain types of chimeras, including humans with transplanted animal material or animal genes. In general, this will not affect the current catalogue of fundamental rights. The right to equality and the right to vote, for example, can be applied normally. This does not mean that it will always be easy to determine whether—and when—a distinction is justified between enhanced and non-enhanced humans, but that is not a new problem. After all, the right to non-discrimination often requires complex and sensitive considerations. We may have to add new fundamental rights to those already existing, such as a right to identity for clones (and other humans), a right to mental integrity for cyborgs and a right to forget or be forgotten.

From a legal-theoretical perspective, human enhancement also means that, in the long term, discussion is required to determine who should be entitled to fundamental rights. This does not concern enhanced humans that evolve from *homo sapiens*—who are, after all, 'normal' human beings—, but it does concern androids and other robots. If in the future they can function in society in a way comparable to natural or legal persons, they will also—whether or not after demonstrations on

Trafalgar Square—be able to claim the legal protection of fundamental rights, in so far as this makes sense depending on the nature of the fundamental right in question. The legal status of frozen bodies or bodily materials—as long as there is a theoretical possibility they might be defrosted in the future and brought to second life—remains a question yet unanswered.

Second of all, fundamental rights also have consequences for the development of human enhancement, and I believe this is the most relevant part of the research question. Fundamental rights comprise a certain right to enhancement. In principle, inviolability of the body implies that citizens have the right to improve their bodies, in ways ranging from plastic surgery to bionic arms and brain implants. With some exaggeration, it can even be argued that the government has a duty to encourage certain forms of enhancement, for example by stimulating scientific research into slowing or stopping the ageing process (improvement of public health) or even into genes that are associated with altruistic characteristics or sensitivity to arts (social and cultural development). Even without embracing the transhumanists' proactive enthusiasm, there is good reason to cherish and stimulate certain aspects of human enhancement—especially those that promote the diversity and richness of human life.

At the same time, however, there is also cause for concern and reticence. Enhancement can be positive and attractive, but it can also lead to citizens feeling pressured to go along with a socially or politically desirable enhancement trend. In my opinion, citizens also have a right to *non*-enhancement: those who want to should be able to remain human, with all their good and bad qualities. If, for example, the ageing process can be slowed down, it remains to be seen whether politicians can oblige citizens not to age for social-economic reasons. In order to counterbalance such social and political pressure, we may want to consider introducing a fundamental right to imperfection, a right to ageing and even a fundamental right to die. It may also be desirable to extend the government's duty of care to include pluralism of humanity, in order to prevent the development of a bleak uniformity of enhanced humans on the macro level which could destroy the richness of human experience.

And on the basis of the conclusion that enhancement can be a right but never an obligation, I believe that our main concern for the future should be to prevent non-enhanced humans from being fundamentally disadvantaged because they do not meet some kind of enhancement standard. Fundamental rights will have to play an important role in preventing 'normal' humans from becoming an underclass to enhanced humans.

London, 28 June 2079, from our reporter
The demonstration of orthodox humans that took place on Trafalgar Square yesterday proceeded peacefully considering the circumstances. Around 20,000 humans, who, for various reasons, refuse to follow the regular enhancement procedures, answered the call of the Human Alliance to demonstrate against their disadvantaged social position. 'The discrimination against us normal people has to stop,' says Andy, a 36-year-old palaeoman from Bristol. 'We have the right to employment, but nobody gives us jobs. Most of us are completely healthy, but we have to pay three times more insurance than genetically

modified humans. There is hardly any up-to-date learning material for our children; everything is based on better-brain-education nowadays.'

Despite the atmosphere of solidarity, the mood remained somewhat resigned. The turnout was disappointing, many members of the HA not being able to afford the journey to London, and the demonstrators were almost completely ignored by the neohumans speeding by. The police did fine several cyborgs for public humiliation after they stopped at the demonstration and—with a rather palaeo-sense of humour—shouted 'Hey, Neanderthals!' at the demonstrators.

A small ray of hope was offered to the palaeohumans by the speech of the Minister of Justice Warwick, cloneson of the colourful UK scientist of the early twenty-first century. He emphasised that society must respect the ethical positions of minority groups and that palaeohumans can still fulfil a useful role in society. He did not, however, take up the HA's 10-step plan, considering affirmative action for government jobs to be a step too far, and a right to compensation for palaeomedical facilities and the plan to stimulate non-brain-interactive cultural shows to be too expensive. He did promise to investigate possibilities to improve employment for palaeohumans and to request the Cabinet to fund learning materials for palaeochildren.

Bibliography

Atwood M (2003) Oryx and Crake. Bloomsbury, London

Calverley DJ (2005) Toward a method for determining the legal status of a conscious machine. In: Proceedings of the symposium on next generation approaches to machine consciousness, pp. 75–84, University of Hertfordshire, Hatfield, UK, AISB, 12–15 Apr 2005

Constitution of the Kingdom of the Netherlands, http://www.rijksoverheid.nl/documenten-en-publicaties/brochures/2008/10/20/the-constitution-of-the-kingdom-of-the-netherlands-2008.html

Fukuyama F (2002) Our posthuman future: consequence of the biotechnology revolution. Farrar, Straus and Giroux, New York

Harris J (2007) Enhancing evolution. Princeton University Press, Princeton

Koops BJ, Hildebrandt M, Jaquet-Chiffelle D-O (2010) Bridging the accountability gap: rights for new entities in the information society? Minn J Law Sci Technol 2:497–561

Matthias A (2007) Automaten als Träger von Rechten. Plädoyer für eine Gesetzänderung. Dissertation, Humboldt Universität, Berlin

Solum LB (1992) Legal personhood for artificial intelligences. N C Law Rev 70:1231–1287

Teubner G (2007) Rights of non-humans? Electronic agents and animals as new actors in politics and law. European University Institute, Florence, Max Weber Lecture no. 2007/04

Chapter 13
Conclusion: The Debate About Human Enhancement

Bert-Jaap Koops

Abstract The social debate surrounding the concept of human enhancement provides a guideline for reflection and further discussion. The debate often centres on definitions: what is a human being, what is an 'ideal' human being and what does it mean to 'enhance' it? A second common issue is a concern when technologies move from medical treatment to enhancement, raising issues of (in)equality and responsibility. Although widely different manifestations of human enhancement are discussed, they constitute a common debate: who should take life-changing decisions, and how should we deal with different frames, ethical outlooks and risk attitudes? This debate is often polarised. We can enhance the quality and effectiveness of the debate, as is attested by the contributions to this volume, by means of clarifying, broadening and deepening the subject matter. If we analyse human enhancement issues in clearer, richer and more nuanced ways, we can take the discussion to a next level.

This book has presented a variety of perspectives on human enhancement. What is fact and what is fiction? Why are we so fascinated by human enhancement and how should we deal with all current and future dilemmas? In this chapter, I will try to pinpoint some common thoughts visible in all perspectives offered in this book as an incentive for further reflection and discussion about these questions. My guideline will be the social debate on human enhancement—a debate which is, like this book, fed by fact, fiction and fascination. What is the debate about, and what

---

Translated by Maartje Wienke.

B.-J. Koops (✉)
Tilburg University, TILT, P.O. Box 90153, 5000 LE, Tilburg, The Netherlands
e-mail: E.J.Koops@uvt.nl
http://www.tilburguniversity.edu/webwijs/show/?uid=e.j.koops

B.-J. Koops et al. (eds.), *Engineering the Human*, DOI: 10.1007/978-3-642-35096-2_13,
© Springer-Verlag Berlin Heidelberg 2013

are its controversial issues? How can the insights from various disciplines contribute to this debate? And how can the debate about human enhancement be brought to a higher level?

An Outline of the Social Debate

The enhancement of human beings is being discussed everywhere and on all levels: in newspapers, on television, in the cinema, in conference centres, in academic journals and books and, last but not least, in politics. In these debates, human enhancement has many manifestations, varying from smart pills, plastic surgery, genetic selection, computer brains, clones and chimeras to completely computer-controlled robots. This book contributes to this social debate by having many authors from a variety of disciplines reflect on human enhancement from various angles. It also offers a reflection on the debate itself by looking at its various aspects: its liveliness, diversity and wealth of visions, yet also its intenseness, its bias and its polarisation. The enhanced human is a creature with many faces, as is the social debate surrounding it. This becomes clear when we try to outline the debate, partly on the basis of the previous chapters.

The debate seems to focus first of all on the intrinsic tension between both elements of the term 'human enhancement'. We feel that the enhanced human being does not fit well with our existing concept of being human, because something 'artificial' is happening—the enhancing—which violates the 'natural' within human beings. 'Ordinary' humans and 'artificial' creatures are therefore often positioned on opposite sides of the debate. Cyborgs and androids, genetically manipulated embryos and clones, athletes with miracle-working prostheses and artists voluntarily undergoing megamorphoses—they are all strange creatures forcing us to acknowledge that 'humans' are changing. And all this despite the fact that we so dearly want to stay human. Here, the debate's tension becomes visible: we do not mind progress, as long as our 'being human' is not affected in its core.

However, a closer look reveals that the contrasts between the opposing views are not as clear-cut as they seem at first sight. What is the definition of a human being? The authors of this book are unable to give one. On the contrary, many of them emphasise that concepts of being human are changeable, in time as well as in place. In the eighteenth century, Linnaeus had no problem aligning Chinese and Hottentots with *homo monstrosus*, while the orang-utan—literally 'forest man'—was classified as *homo sapiens* (Chap. 2). We have a different view nowadays.

Also changing are views of who are seen as 'ideal' human beings. Advertisers at the end of the nineteenth century enticed ladies with '"Fat-ten-U" Food to Get Plump', while the women pictured in those advertisements (Chap. 3) would nowadays be regarded as obesity patients. According to various authors, the line between beautiful and ordinary, between illness and personal characteristics, between curing and enhancement is wafer thin and changeable.

13 Conclusion. The Debate about Human Enhancement

Something similar can be said of the term 'enhanced'. The artificial with which this term is often associated in the debate is connected with the state-of-the-art in technology and is therefore as much related to time and place as is the definition of human beings. Habituation and the spread of new technologies play a crucial role in how artificiality is experienced. While Louise Brown as the first IVF baby was the prime example of a 'made' human being in 1978, IVF children today are seen as rather normal. The boundary of 'unnatural reproduction' seems now to have shifted to embryo selection, where choosing between different embryos based on genetic knowledge instead of IVF itself is seen as human enhancement (Chap. 9). It is quite possible that in another 30 years or so embryo selection will be very common, while genetic manipulation of embryos will then be at the centre of the debate about human enhancement. In addition, the topical question of whether we can and want to create synthetic life might also be a completely common and accepted practice in 30 years (Chap. 6).

This changeability of the concept of being human and of technology has as a consequence that what we classified as 'enhanced' human beings yesterday seem to be perfectly normal human beings today. The enhanced human being from current future scenarios might be seen as a perfectly normal person when today's future has become the present. This calls for humility in the debate. The 'natural' which would be contaminated by a new form of 'enhancement' is dependent on time and technology. Radical changes and vital differences between the 'ordinary' and the 'enhanced' human being are often stressed too much; various authors point out that there are no earthquakes happening, that differences are gradual, and that changes happen in stages. What is more, the enhanced human being in fact fits very well within the evolution of mankind (Chap. 8). Yet debates exist because of differences and they prosper on discontinuities. Therefore, we cannot and should not close our eyes to the fact that many participants in this debate are genuinely concerned about the changes in being human caused by ongoing technological developments. The field of tension between 'enhancement' and 'human' is continually shaped by debates about the boundaries of 'being human', which, in today's experience, are under pressure.

Another main thought is that many developments surrounding human enhancement take place in an area where the medical sector diffuses into other fields of application. Most technologies facilitating enhancement, such as biomaterials, tissue engineering, chip implants and genetic techniques, are developed for medical purposes. Although the boundaries of medicine are not always clear, *within* this area technological developments are often undisputed. However, discussion arises when these technologies are used or threaten to be used for different purposes. For which genetic characteristics do we want to permit embryo selection? Are athletes allowed to receive prostheses which enhance their performance 'more than normally'? Can an 'ADHD pill' also be used to increase a 'normal' person's concentration? The transition from curing (making better) to enhancing (making better *still*) turns out to be a crucial boundary in the debate.

One of the most important points raised by crossing this boundary is whether inequality is legitimised. Human enhancement is often associated with stronger, more beautiful and/or better humans. This gives rise to the question of whether a gap between enhanced and non-enhanced human beings is emerging. While the prohibition of discrimination restricts the 'favouring' of enhanced human beings, it also poses the question of when a *distinction is justified* (Chap. 12). After all, humans are not equal and their equality (or inequality) has to be determined again and again in each specific context. The subject of inequality returns in a different manner in issues surrounding the individual choices made by humans regarding their own lives (like the artist Orlan, Chap. 3) or the lives of their children (Chap. 9) and the liability that does or does not come with making these decisions (Chap. 11). Here, an opposite spectre features in the debate about human enhancement: the spectre of homogeneity. Are we in danger of converging to a uniform 'ideal type' of human being which leaves out any form of abnormality because of enhancement techniques? The leap in the dark or accepting the unknown—consciously *not* pushing future human beings in a certain direction—should remain possible (Chap. 11). It is remarkable in this respect that opting for 'non-enhancement'—also when medical decisions are concerned—is particularly respected when this choice is rooted in religious beliefs, as is the case with not vaccinating newborn children (Chap. 11).

Inequality is one of the normative questions that play a part in the debate surrounding human enhancement. Other normative issues are, for instance, physical integrity, social fundamental rights and ageing (Chap. 12). A continually recurring question raised by these normative issues concerns individuals and their responsibilities. Exactly who makes the choices regarding human enhancement? Do individuals decide or do their parents decide? What responsibility do health professionals have in these matters? Or are the choices made by politics or the legislature? Is it scientists, who, either consciously or unconsciously, extend technological possibilities over time? Or is it society as a whole with its weapon of social pressure? It turns out that choices are made on various levels by all sorts of actors (Chap. 9). This raises questions concerning the actual or desired division of responsibilities: who is in the best position to make rational and informed choices about human enhancement? The debate, as illustrated in several of the contributions to this book, demonstrates that there are no easy or unambiguous answers to these questions.

One of the complicating factors is that the debate *is* not always rational and informed. Discussions on human enhancement use many examples from fiction and fantasy, from stories and myths. Views are not always based on scientific insights into the current state-of-the-art in technology and realistic, scientific estimates of future developments surrounding new technologies, but on perceptions and expectations of these technologies as translated and interpreted—to use a term from communication sciences: mediated—in stories from popular media or fiction. The authors of this volume themselves frequently refer to stories from film or literature, including *Frankenstein*, *Brave New World*, *The Matrix* and *Gattaca*. A mythology of human enhancement is an unavoidable fact. We need stories and myths to give meaning to the world around us, especially when that world is becoming increasingly complex and difficult to understand in light of new

13 Conclusion. The Debate about Human Enhancement

technologies. Fiction offers an opportunity to anticipate and to reflect on unknown futures.

At the same time, we have to realise that using myths and stories in the debate creates a gap between those thinking about the consequences of technology on the one hand (citizens, policy makers and social scholars) and those developing these technologies on the other. Natural scientists often do not recognise themselves in the picture painted by the debate of 'their' technology, which may be one of the reasons they sometimes seem to participate in the social debate less than scholars from the humanities and social sciences. Consequently, fantasies gravely distorting reality are barely or not at all corrected, and myths can start leading a life of their own. This is the case in reproductive cloning, for instance, where several non-representative spectres from literary fiction seem to form the dominant image in the debate and in public policy (Chap. 10). Here, the mythology of human enhancement appears to be a vicious circle that is difficult to break.

A Reflection on Human Enhancement

After this overview of the social debate I would like to submit human enhancement and the debate surrounding it to a closer analysis, using comments, refinements and insights offered by the different contributions to this book. First of all, questions are raised by the great diversity of manifestations. The concept of human enhancement is hard to define (Chap. 2), and it is even questionable whether one can speak of a single debate. Are we not actually talking about different phenomena when we consider cyborgs, artificial intelligence, body artists, DNA research and embryo selection? The use of 'human enhancement' as an umbrella term does not always seem to fit, and various authors struggle with positioning their topic and perspective within this volume. At the same time, we see that the 'enhanced human being' keeps returning, although sometimes in different forms, and that all forms show, as Wittgenstein calls them, family resemblances. Cyborgs, robot brains, selected embryos and Dutch ambassador of plastic surgery Marijke Helwegen are no clones of each other, yet they are metaphorical brothers and sisters, with an occasional distant aunt.

The questions concerning these various forms asked in social debates all have something to do with the earlier-signalled tension between enhancing, which is experienced as 'artificial', and being human, which is seen as 'natural'; a certain action interferes with being human in a different or more profound manner than what is regarded 'normal'. This is why I deem it useful to continue speaking of 'the' debate on human enhancement, with the debate's borders shifting in place and time with technological and social developments. My firm conviction is that a reflection on this debate is useful and productive, as it can teach us something about what it means to be a human being in this day and age where technology is used in all sorts of manners to direct or adjust human development.

The next step is to draw various conclusions. Only a few decades ago, social engineering was an important topic for debate; nowadays we talk about 'human engineering' (Chap. 4). This does not mean, however, that all emphasis should be on the individual. Some debates take place on the micro level of individual choices and personal responsibilities, yet other debates about human enhancement take place on the macro level of political choices and social pressure. In the last instance, human enhancement still contains elements of the concept of 'malleable society'. How do we arrange society in such a manner that humans can develop themselves in the best possible way? However, micro and macro levels are not always clearly distinguished in the debate. As these two mutually shape each other, a clear separation between the levels is not always possible. For instance, legal responsibility for individual choices depends on a norm set by society (Chap. 11), and the political choice of when embryo selection should or should not be allowed has obvious consequences for the individual choices of would-be parents (Chap. 9).

Still it is desirable to clearly distinguish between the two levels, as the main question of how far we want to go with the enhancement of human beings is formulated differently on the different levels. On the micro level of individual decisions, this question leads to issues about boundaries of individual freedom of choice, where factors such as autonomy and fundamental rights are prominent. On the macro level of policy decisions, the central question asks when and how public interest is affected by developments in the field of enhancement and whether and how these developments can be steered in certain directions.

The phenomenon of 'framing' plays an important role in this respect: directing a debate towards a certain course by choosing a frame that evokes certain formulations and metaphors. Someone who starts a debate by asking what fundamental rights cyborgs have or should have (Chap. 12) does not only position the debate at the macro level but also—consciously or unconsciously—implies that the legislature may have an important role to play in guiding the development of cyborgs by means of fundamental rights. This could lead to the implicit basic assumption that we must be reticent in the development of cyborgs—until proven otherwise. However, when a debate starts with the question of whether someone has the right to implant technology in order to enhance themselves even further (as argued by Kevin Warwick, Chap. 7), the debate is explicitly positioned on a micro level. This then is associated with individual autonomy and freedom of choice, and can lead to the implicit assumption that we should not interfere with the development of cyborgs—until proven otherwise. In other words, framing the debate through the question asked, for instance by positioning it on a micro or a macro level, influences the debate's implicit basic assumptions.

The next observation is related to this. Participants in the debate use different strategies for handling the uncertainty accompanying the various forms of human enhancement. Many experience these uncertainties as threatening and formulate questions concerning human enhancement in terms of risk regulation. How can we protect humanity from becoming an underclass to a new class of 'post-human beings'? How can we ensure that we, despite all these enhancement techniques, still remain human? Others see this uncertainty as offering opportunities; they

13 Conclusion. The Debate about Human Enhancement

formulate questions regarding human enhancement as challenges to become better human beings (or post-human beings). Do we not have a moral duty to improve human life if new technologies allow us to do so? Why should we not take the leap in the dark and become a new species, *homo manufactus*? I think that these kinds of stands are not so much related to the field in which one works—both cyberneticist Kevin Warwick and ethicist John Harris (Chap. 7) are examples of the second category—as they are to an individual's risk-avoiding or risk-seeking personality and their worldview or religion. These last factors are particularly important as they colour one's view on what makes humans human.

A parallel observation is that the debate routes along different ethical positions. Some approaches rely heavily on utilitarianism—a cost–benefit analysis of what provides most utility, such as the health-economic approach to embryo screening (Chap. 9). Others apply a human rights view, which is based on fundamental rights and liberties of citizens as recorded in various human rights acts (Chap. 12). Still other views take as their starting point the concept of human dignity, which plays an important role in the worldwide ban on reproductive cloning (Chap. 10). Such views are based on the fundamental assumption that human dignity—a person's intrinsic value as a human being—must never be violated. However, the definition of this intrinsic value is often unclear; it could for instance be related to a certain religious concept of being human or to a Kantian idea of ethics. As Roger Brownsword has demonstrated in his analyses of the regulation of biotechnology (see for instance *Rights, Regulation, and the Technological Revolution*, 2008), the bioethical triangle of utilitarianism, human rights and 'dignitarianism' is a complicating factor in the decision-making process regarding human enhancement issues, since the fundamentally different starting points are difficult to combine. What may be convincing arguments for one group may not catch on with the other, regularly causing debaters to misunderstand each other.

Moreover, some views seem to not receive as much attention as others. As pointed out before, natural scientists are rather silent in the social debate (or perhaps they are silenced by the myth-eager media?), even though they are able to reduce future-oriented and sometimes speculative debates on human enhancement to realistic proportions. Furthermore, the debate is dominated by a single view, which concentrates on new technologies that enhance human beings in one way or another. Because of this dominant view, some family members of human enhancement are pushed back, while they do in fact have something to add to the debate, as they raise the same questions. Not only new technologies facilitate human enhancement; there are additional instruments to steer human development and behaviour: education and awareness raising, social norms and social pressure and various preventative measures influence the future of human beings as well. This is why, for instance, embryo selection—a preventative procedure which does not literally 'enhance' human beings—belongs to this category, along with genetic manipulation and cloning. After all, they raise similar questions regarding ethical admissibility, responsibility and liability concerning reproductive decisions. The social pressure connected with beauty ideals has more influence on human enhancement than the fact that technology enables plastic surgery. Here, the debate should therefore concentrate mainly on the role played by beauty ideals in society rather than on the technology of plastic surgery.

Another underexposed aspect is that the debate does not always have to be about 'enhancement'; the Dutch term '*maakbare mens*' (literally: makeable man) is in this respect better suited than the English 'human enhancement'. Actions that intervene in our being human are not necessarily intended to beautify, enhance or strengthen human beings. Sometimes the opposite is the case, as with the lesbian couple who desired a deaf child because both of them belonged to the deaf community (Chap. 12) or the *nullo*, the man who had his genitals removed (Chap. 3). When discussing the boundaries of human 'enhancement', we should also, or in particular, discuss the boundaries of human 'degradation'. And as history has shown (Chap. 3), enhancement techniques are not only used to become different (either better or worse) but also to become normal or ordinary, in order to not stand out. Here too, the boundaries of both sides of the topic should be discussed. Limits to differences between 'enhanced' and 'non-enhanced' human beings (to prevent a social gap) should be discussed as well as limits to the 'normalisation' of human beings (to prevent social homogeneity). Although this last aspect is mentioned occasionally, in my opinion it should receive more attention.

A last observation concerns the fairly polarised state of the debate on human enhancement. Debaters are either convinced advocates of a certain form of enhancement or else convinced opponents. Opinions are based on diverse world views and ethical movements and are partly founded on perceptions and stories that diverge considerably from scientific reality, causing misunderstanding among debaters. Clearly, this does not enhance the quality and effectiveness of the debate. Can measures be taken to remedy this situation? I believe so.

A Step Further

Although human enhancement and the surrounding debate are both multifaceted and complex, I think this book can further the debate in three ways. The first of these concerns sharpening and clarification. A clear definition of 'human enhancement', either of the term as a whole or of its elements 'human' and 'enhancement', does not exist, and there is no point in trying to develop uniformly accepted definitions, as the concepts are too complex and contested to do so. We can have a meaningful conversation about this phenomenon even without clear definitions, as this book has hopefully shown. However, it *is* useful to keep clarifying in a debate what it is we are talking about. What exactly is the 'enhancement' that is being discussed, and how does it influence 'being human'? Is it fundamentally different from existing forms of influencing human beings? Or, in other words, is it a true discontinuity or is it a gradual process? And on the basis of which principles do we decide this?

Another important point of interest concerns the sources from which examples and information are drawn, especially with regard to technological developments.

13 Conclusion. The Debate about Human Enhancement

Do they spring from imagination and stories in fiction or futures scenarios or from scientific sources? Are we talking about forms of enhancement that are present today—experimental or not—or about medium- or long-term expectations, and what are they based on? Is the research debated still in its infancy or is the debate based on technology with a reasonable reservoir of empirical knowledge? In short, there is much to be gained in trying to better distinguish fact from fiction.

The last and possibly the most important point in relation to clarification is to make implicit assumptions explicit—for example concerning someone's concept of being human, ethical basis and worldview. Debaters can try to make clear, to themselves and to others, why they hold certain views and why they deem certain arguments more important than others. This does not necessarily bring the different points of view closer together, but it does make it easier to pinpoint where exactly the differences lie.

The second way to enhance the debate is to *broaden* the discussion. As this collection of essays has shown, scholars look at human enhancement from various angles; they each discuss a fragment of the concept, from their own background and language. The confrontation of all these perspectives, as we have strived for in this book, helps everyone to get a clearer insight into their own assumptions and the frame of mind from which problems are approached. Sometimes, this helps to stretch that frame of mind to gain a different perspective on the problem, which can help in finding new solutions. I am convinced that fundamental questions concerning human enhancement can only be answered meaningfully when different disciplines are combined, when the humanities and social sciences on the one hand and the natural, bio- and neurosciences on the other enter into a joint discussion. Such a multidisciplinary approach will not immediately provide answers, but it will help to ask the right questions—questions that will not surface if a monodisciplinary approach is applied. After all, someone who only has a hammer and nails in her toolbox is unlikely to ask what is the best way to saw a plank. By combining the toolboxes of different disciplines it becomes possible to map the problem, after which possible solutions can be explored.

Another way to broaden the discussion and thus gain a better understanding of the subject is to put it into a historical perspective. The realisation that human beings have been concerned with enhancement techniques and enhancement ideals for centuries (Chaps. 2 and 3) can help to calm down some current discussions—some forms of enhancement are not *that* radical and far-reaching. Still more important is the realisation that human enhancement is in fact a continuous project of *homo sapiens*, and that using new technological and social possibilities to change ourselves is perhaps one of those characteristics that make humans human. Applying a historical perspective eases the tension between 'artificial' enhancement and 'natural' human beings by showing the relativity of both concepts. Human beings are artificial creatures by nature.

A third opportunity to further the discussion is to try and have a more *in-depth* debate. While analysing the debate, I signalled several aspects that were insufficiently highlighted, such as the non-technological variants of enhancement, variants that are aimed at deterioration or normalisation instead of enhancement, and the significance of worldviews, religious or otherwise. By underestimating such aspects we tend to leave certain choices concerning enhancement out of the normative assessment. Paying more attention to these aspects can lead to a more in-depth debate, as it helps to sharpen relevant questions. This goal can also be achieved by moving beyond the false oppositions which sometimes dominate the debate. Nature versus culture, evolution versus enhancement and created human beings versus enhanced human beings—while these are often-heard contrasts, they are generally not at the heart of human enhancement.

The key questions are often to be found elsewhere. What do we want human beings to look like in 20, 50, or 100 years and how can we determine this? How much importance should be attached to worldviews with regard to normative choices on the micro or macro level? Do we want to and can we influence the developments, and if so, who can or should do so in what way? Who will decide on the boundaries of human enhancement in the future? To what extent should and can individual autonomy be secured? When answering these questions, it is useless to presume an unchanged view of humanity which has to safeguard the 'essence' of being human. Nor is it productive to approach new human-changing technologies in a tendentious way, either as threatening or as irresistible.

An in-depth discussion looks for nuances. Technologies are no more developed in a vacuum than are social, legal and ethical norms: they develop in interaction with each other. Instead of techno-determinism (technology itself actively creates new norms) or techno-naiveté (technology passively follows existing norms), we should rather approach technological developments and social norms as a process of mutual shaping. Questions concerning human enhancement operate within the 'co-evolution' of technology and regulation, and we can only approach those questions if we are sufficiently aware of the complex interaction between 'enhancing' human beings and judging what kind of 'human beings' we want to be. This interaction is present at all levels and in all places: in the media, in academia, in hospitals, in politics and in pubs. Together we are creating human beings, not least by the debate we are holding on human enhancement.

We cannot make the debate any easier. What we can do, though, is make it clearer, richer and more nuanced, which ultimately makes it more effective.

Index

Page numbers followed by '*f*' indicate figures and '*t*' indicate tables respectively.

0–9
'2010 in Sight' (Schnabel), 47, 58

A
A.I. (Artificial Intelligence), 84
ADHD, 114, 176
Ageing, 36–37
    Fat-Ten-U, 38, 39*f*
    'puberty gland' dubbing, 37
Albinism, 176
All Souls' Day (Nooteboom), 29
Androids, 83, 84, 99, 168
    walking android, 90*f*
Anti-ageing, 3, 4, 7
    absence of illness, 174, 175
Anthropomorphic Creatures, 23*f*
Appeal to Human Dignity (AHD), 166
Artificial humans from past, 19–22
Artificial intelligence, 20
    applied to human beings, 3, 5, 7
    and human cognition, 88–89
    in intelligent computer construction, 89
    threat to humankind, 88
Artificial lenses, 115
Artificial man, 3
    examples, 13–14
    magic, 19
Artificiality, 25
Asimo robot, by Honda, 90*f*
Asimov, 84, 85

B
Bio-adapter, 26
Bio-bricks, 77
    and Lego bricks, 77

Biodegradable suture thread, 74
Biological thinking, 46, 49
'Biologism'. *See* 'Biological thinking'
Biomaterials, 73–74
    point mutations, 73
    protein reproduction, 73
    recombinant DNA techniques, 73
Body
    as environment, 107–108
    as mask of death, 31
    as work of art, 40–41
Body images, 37–40
Body movement factor, 95
Bokanovsky process, 140
Boys from Brazil, The (Levin), 115, 130,
        132, 133
BrainGain, 86–87
    'Brain Computer Interfacing' (BCI), 86
    and second Warwick experiment, 87
Brave New World (Huxley), 17, 46, 115, 130,
        132, 142
    inferior clones, 142
Breeding humans, 14–17
    artificial, 16
    ethics, 17
    self-direct human evolution, 16
    social Darwinism, 16
    tampering natural forms, 14
    well-bred, 15

C
Camper, Petrus, 33
Can Man Be Modified? (Rostand), 8
Case Against Perfection, The, 6
Charter of Fundamental Rights of the European
        Union ('European Charter'), 168

194   Index

Chastisement, 31
Chimera, 168
Chorionic villus sampling (CVS), 154
Christianity, making man, 12
City of the Sun, The (Tommaso Campanella),
    15
Clinical trial, 124
Clone, 168
Clone (Cowper), 132
Clone gap, and mass production, 140–144
Cloned Lives (Sargent), 133, 134
Cloning and identity, 131–132
Cloning of Joanna May, The (Weldon), 130,
    132, 135
Coeliac disease, 121
Colour-blindness, 176
Complex regional pain syndrome (CRPS),
    156–157
Concept of Passing, 35
Concept of being human, changeability of, 185
Conditioning people, 17–19
    collective education, 19
    nature-versus-nurture debate, 18
Consumer robots, 90
Convention on Cybercrime
    art. 19, para. 2, 172
Copy of unique original, 133–134
    as reason for cloning, 133
Cosmetic surgery, history of, 35
    reconstructive surgery in Nazi
        Germany, 36
    mimicking parents, 35–36
    social function of imitation, 36
    similarity, 36
    war injuries, 36
CPB (Netherlands Bureau for
        Economic Policy Analysis), 47
Creating humans, 79–80
Creating life, 77–78
Curious scientist, 144–145
Cybernetics, 2
Cyborg, 168
Cystic fibrosis, 65, 121

D
Dancing and jumping robots, 93, 94
David, robot, 84, 138
'De maakbare mens' ('makeable man'), 3
    diverse connotation of, 3
Deafness, genetically induced, 176
Defending the Genetic Supermarket
        (Gavaghan), 5–6
Golem, Der (Meyrinck), 20

Design phase, 49
Designer babies, 49
Dexter (Anybots), 93
Diagnoses of the present, 48
Die Perfektionierung des Menschen
        (Gesang), 3, 6, 27
Die Verbesserung von Mitteleuropa, 26
Disciplining behaviour, 52–53
    have-nots, 53
    haves, 53
    'Hyper-individualism', 52
Displacement of politics, 126–127
    genetic research promises, 126
    QALYs, 126
Diversity in unity, 134–135
    difference from donors, 136
    individual preferences, 135
Doctrine of the pure heart, 31
Down's Syndrome, 67, 117
Duchenne muscular dystrophy (DMD), 121
Dutch Constitution
    Article 1, 169
    Article 3, 170
    Article 4, 170
    Article 10, 171
    Article 11, 171
    Article 13, 172
    Article 22 Section 1, 173
Dutch Health Council, 117
    screening definition, 116
Dutch Huntington Society, 124
Dutch Supreme Court, 157
    immunity of parents' choice, 158

E
E. coli bacteria, 75
Elasticity, of protein, 75
Elementary Particles, The (Houellebecq), 46,
    58
Embryo splitting, 131, 140, 144
Engineerable and Self-mutating Human,
    The (Hendriks), 49
    types of enhancement, 49
Enhanced humans, 95–96, 167
    and human rights, 179–180
    socially desirable, 96
    technologically feasible, 96
Enhancement of child, 151–153
    examples, 153–155
        amniotic fluid test, 153–154
    parents duty, 157–158
Enticing behaviour, 53–54
Entities with legal personhood, 168

Index 195

legal person, 168
natural person, 168
Essayistic reflections, 48
Eugenics, 16
European Charter, 168, 175
   Article 3(2) (d), 172
European Convention on Human Rights and
   Fundamental Freedoms, 168
   Article 8
European Convention on Human Rights, 171
Evolution and environment, 106
   niche construction, 106
Evolution and humans, 105–106
   mutations, 105
Evolution and life expectancy, 103–104
   Darwinian fitness, 104
   extrinsic mortality, 104
   'how' question, 104
   intrinsic mortality, 105
   trade-off theory, 104
   'why' question, 104
Experience 2030 (Essent), 51
   haves and have-nots, 52
Extreme body modification, 40
Extrinsic mortality, 104

F

Fabricated Man (Ramsey), 8
Familial hypercholesterolemia, 119, 120
Forest satyrs, 23*f*
Free choice
   future of, 161
      civil law, 161
      inoculation, 162
      liability law, 161
      unusual choices, 162
   influencing free choice, 160–161
      insurance companies, 160
      legislature, 160
   of parents, 158–159
      limitations, 159
Fundamental rights, 165–182
   Convention on the Rights of the Child,
      Article 8, 174
   Dutch Constitution
      Article 1, 169
      Article 3, 170
      Article 4, 170
      Article 8, 174
      Article 10, 171
      Article 10, Sections 2 and 3, 174
      Article 11, 171
      Article 13, 172

      Article 22, Section 1, 173
   European Charter
      Article 3, 175
      Article 10, 175
      Article 11, 175
   International Covenant on Civil
      and Political Rights, 168
   International Covenant on Economic,
      Social and Cultural Rights, 168
   homo politicus perfectus, 171
   legal persons and animals as
      holders of, 178–179
Future of the Medical Doctor,
   The (Groenewegen, Hansen
   and Ter Bekke), 55, 58

G

Gattaca, 63
   genetic makeup, 64–65
   genetic selection, 67–68
   genetic susceptiblity, 66–67
   hereditary disorders, 65–66
   highly heritable, 66–67
   individual's genetic profile, 114
      and social inequalities, 114
Genetic enhancement
   in future, 69–70
Genetically induced deafness, 176
Genomics 2030 (de Graef and Verrips), 54, 58
Geulincx, Arnold, 13
Good's entry into Ghent, 32
Gluten intolerance. *See* Coeliac disease
Groundhog Day, 176
Gulliver's Travels, 71, 80
Guthrie test, 117
   concept of 'enhancement', 122
   in newborns, 120
   Tandem Mass Spectrometry, 121

H

'Haves and Have-nots' scenario, 53
High Energy, 54
Historical situation, novelties, 25–27
Hollywood film, 63
Homme machine, 20
Homo monstrosus, 22, 185
Homo politicus perfectus, 171
Homo sapiens, 22, 185
Homo Sapiens 2.0: Festival about the
   'Makeable Man', 4
Homosexuality, 176
Homunculus, manufacturing, 21*f*

Human, 168
Human cloning, 129
  recreate oneself, 130
  reproductive cloning, 131
  therapeutic cloning, 131
Human enhancement, 3, 42–43, 103, 167–169
  cause for concern and reticence, 181
  cost–benefit analysis, 189
  development of human
    enhancement, 180–181
  embryo selection, 176
  through evolution, 107–108
    genetic and environmental
      manipulation, 108
    life expectancy of 70-year-old Swedish
      men, 108*f*
  and existing fundamental rights, 169–174
  'favouring' of enhanced
    human beings, 186
  fiction and fascination, 183, 186
  fundamental rights, 167, 181
    non-'natural' additions, 167
  future
    broadening research, 191
    'co-evolution' of technology
      and regulation, 192
    in-depth debate, 191–192
    understanding subject, 191
  genetic manipulation, 176
  inequality and, 186
  legal questions concerning, 154
    amniotic fluid test, 154
    choices made after birth, 155
    chorionic villus sampling (CVS), 154
    spina bifida, 154
  limits of, 55–58
  maakbare mens, 190
  and new fundamental rights, 174–178
  normal human, 181
  orthodox humans, 181
  overview of, 56–57*t*
    framing, 188
  and prevention, 115
  reflection on, 58–59, 187–188
    artificial and normal human beings, 187
    human engineering, 188
  right to imperfection, 176
  social debate, outline of, 184–185
  'ordinary' humans and 'artificial'
    creatures, 185
Human Enhancement (Savulescu
  and Bostrom), 6
Human enhancement and genetic
  testing, 113–114

economic decision, and setting priorities,
  117–118
  cost-effectiveness analysis, 118
  quality-adjusted life years
    (QALYs), 118
  transparent decision-making
    process, 118
  social–psychological decision, vision
    of parents, 120–122
  child's future development,
    knowledge, 121
  cost-effectiveness considerations, 120
  decision aids, 122
  Guthrie test, 120
Human interface, 91
  iCat project, 91
  Kismet robot, 91
Human life expectancy, perspective, 102–103
Human lifespan extension, and side
  effects, 108–109
  postponing start of family, 108–109
Human machines and mechanical
  humans, 98–99
Human rights, *see* Fundamental rights
Human substitute robots, 94–96
Humanoid cyborgs, 85
Human-oriented society, 85
Human–Robot Relationships, 84
Humans and nonhumans, 22–25
Huntington's disease, 124, 176
  development of genetic test, 124
Huxley, Thomas, 17
Hyper-individualism, 42, 52
  in *Preview 2030* (Essent), 58

I

IBM's computer Deep Blue, victory over chess
  grandmaster Garry Kasparov, 89
iCat by Philips, 93*f*
Idem-identity, 131
  and ipse-identity, 132
Identical twins, 130
Immortality, evolving towards, 109–110
  grandmother effect, 111
  lifespan extension
    menopause, 110
    and procreation, 110
Imperial Earth (Clarke), 133, 138
Incidental death. *See* 'Extrinsic mortality'
Individual autonomy, 32–34
Individualised enhancement, 58
Innocent III, Pope, 30, 41

Index   197

Inside the Domestic Sphere (Koops et al.), 53, 58
Intelligent computers, construction, 89–91
   Asimo robot by Honda, 90*f*
Intrinsic mortality, 105
Invasion of the Body Snatchers, The, 130
Ipse-identity, 132
Ishiguro's mechanical doppelgänger, 94, 94*f*
Island of Doctor Moreau, The (Wells), 46
IVF babies, 166

J

Johenneken mit den Bellen, 32
Joseph, Jacques, 34–35

K

Kidneys, and regenerative medicine, 49, 78

L

L'imparfait du présent (Finkielkraut), 151
Lamarckian inheritance, 19
League table, example of, 118*t*
Lebensborn, 16
Left-handedness, 176
Legal concepts, definition, 156–157
   civil law, 156
   criminal law, 156
Long-term studies, 48
Loyal Subject, The (Mann), 34
Lysenkoism (1930–1950). *See* Lamarckian Inheritance

M

Maastricht's bachelor course, 4
'Making man', traditional ways of, 12–14
   artificial methods of production, 13
   human action, 13
Making the Body Beautiful: A Cultural History of Aesthetic Surgery (Gilman), 32
Man in Progress: the Body as a Building Kit (Dorrestein et al.), 46
'Man on the Throne of God?' (van Steenbergen), 48
Man–machine interactions, 95
Man-made man, 3–5
   interventions, types of, 4
   'makeable man', 3
Marcuse, Herbert, 14
'Marketing in Times of Growth', 53
'Marketing in Times of Survival', 53, 54

Mass production and clone gap, 140
Matrix, The (1999), 97
Mind over matter, 40
Minimal genome, 77
MIT's Kismet robot, 92*f*
Mooiemensen.com, 30
Mucoviscidosis. *See* Cystic fibrosis

N

Nanofactories, in cell, 75–76
   cloaking, 76
   gene therapy versus protein therapy, 76
   nanocapsules, 76
   smart drug delivery, 75
   targeting device, 76
Nasal index, 33
Nasen-Joseph ('Joseph of the noses'), 34
National Health Service (NHS), 119
National Institute for Clinical Evidence (NICE), 119
Nature-versus-nurture debate, 18
Need for offspring, 138–139
   clones and donors, 139
   individual creative identity, 139
   reproducing extraordinary individuals, 138
Never Let Me Go (Ishiguro), 130, 132, 140, 141
   inferior clones, 142
New Atlantis, The, 72
'New Human Being in a Future World Society, The', 48
Non-governmental organisations (NGOs), 126
Nose job. *See* Rhinoplasty
NRC Handelsblad, 26, 41, 84

O

Obesity, 176, 184
Old Testament, 32
On the Dignity of Man (Pico), 32
Origin of Species, The (Darwin), 15
Original sin, 18
Oryx and Crake (Atwood), 46

P

Palaeohumans, 25
Parent-child bond, 121
Part human and part machine, 85–86
Perfecting existing humans, 26
Phenylketonuria (PKU), 66, 121
Philosophical Investigations (Wittgenstein), 18

Playing God, 12
Polar Bear Plague on the Veluwe (in 't Veld and van der Veen), 48, 49
Possibility of an Island, The (Houellebecq), 46
Postnatal screening, 158
    Guthrie test, 117, 120
Pre-implantation genetic diagnostics (PGD), 67
    for hereditary forms of breast cancer, 67–68
Pretty v. United Kingdom (2002), 177
Preventive genetic technology, 116
    diagnostics, 116
        obstetricsonography, 116, 117
        pre-implantation genetic diagnostics (PGD), 116
        See also Pre-implantation genetic diagnostics (PGD)
    political decisions, 126
    screening, 116
    to detect potential risks, 116
Preview 2030 (Essent), 52
Professor Williams' products, 38
Promising Technology (van Lente), 59

Q

QRIO robot, 93
Quality and Future (RIVM), 57
Quality-adjusted life years (QALYs), 118
    definition, 118
    league table, 118
Quasi-human, 168
    discrimination against, 168

R

Rapports du Physique et du moral de l'homme (Cabanis), 18
Red-headedness, 176
Reprogenetics, 52
Republic (Plato), 15
Rhinoplasty, 34
Rights, Regulation, and the Technological Revolution (Brownsword), 189
RIVM (National Institute for Public Health and the Environment), 47

S

Selfish Gene, The (Dawkins), 24–25
Service robots, 89, 90, 91
Smart pills, 50–51
    check for deficiencies, 51

    detecting 'anomalies', 51
Smart Pills (van Santen, Khoe and Vermeer), 50, 58
Social bonding, 36
Somatic cell nuclear transfer (SCNT), 131
Spencer, Herbert, 17
Spider silk, 74
    biosynthetic spider silk, 74
Spina bifida, 67, 117, 154
Star Trek's 'Borg', 84
Star Wars, 2, 24
Strength, of protein, 75
Summus Artifex, 12
Systema Naturae (Linnaeus), 22

T

Tagliacozzi, Gaspare, 32–33
    syphilis epidemic, 33
Tandem Mass Spectrometry, 121
Technological prophecy, 27
Technological revolution, 48–50
Trade-off theory, 104
Transhuman, 3
Transhuman League, 165
Transhumanist Declaration, 1–2
    debate, core of, 5–7
    between fiction and fascination, 7–8
    man-made man, 3–5
Transhumanist Party, 172
Transhumanists, 26–27
Turing test, 20

U

Übermensch, 20
UNESCO's Universal Declaration on the Human Genome and Human Rights (1997), 130
Unique (Allen-Gray), 133–134, 146
    clone, closed future, 143

V

Virtual health agent, 55
Vision of the Future, A (Idenburg), 58
Voronoff, Serge Avramovitch, 37
'Voyage to Laputa' (Swift), 71–72
Warwick and BrainGain, 87–89
    compensating for disabilities, 87
    Focus, 88
    improvement or enhancement of unimpaired human capabilities, 87
    standalone machines, 88

Index

W
Welfare and Environment, 57
What Sort of People Should There Be?, 6
Where Idem-Identity meets Ipse-Identity (Hildebrandt et al.), 131
Where Late the Sweet Birds Sang (Wilhelm), 133, 138
Wittgenstein, Ludwig, 18
World of Yesterday, The (Zweig), 37

Y
YouTube, 93
Young Academy, The (De Jonge Akademie) The Royal Netherlands Academy of Arts and Sciences, 8, 73

Z
Zimmer frame, 38

Lightning Source UK Ltd.
Milton Keynes UK
UKOW06n1316020516

273383UK00001B/63/P